What readers say

"Any pirate ship having Richard aboard already has a very remarkable and valuable prize. Having seen his skill with a bow and a flintlock musket I now see he has brains and hidden skills in addition. The author reveals in this book a hitherto unknown facet of pirate lore but the book will be of greater interest not just to pirate fans but anyone who is interested in maritime history. "

"The author has the knack of taking a difficult or complicated historical subject and turning it into an adventure story ... excellently described."

"Real ships, real men ... I'm beginning to see that the *reality* of Pirates as presented by the author in his 'living history' roles is more entertaining than most of the fiction, having far more to offer than *swash-buckling.* A potentially complicated subject matter here made easy to understand ... an education to read and the photographs are marvellous."

"*Interesting, informative, educational and entertaining* ... I'll never look at the sea or stars in the same way again. "

Books by the same author :

On the Robin Hood Trail
in Nottingham and Sherwood Forest

On the Robin Hood Trail ... Again !
in North Notts, Derbyshire and Yorkshire

The World of William Spry, Esquire
A Recreated 18th Century Man

The Pirate Round

Early Eighteenth Century
Maritime Navigation during
The Golden Age of Piracy

Richard Rutherford-Moore

HERITAGE BOOKS
2007

HERITAGE BOOKS
AN IMPRINT OF HERITAGE BOOKS, INC.

Published 2007 by
HERITAGE BOOKS, INC.
Publishing Division
65 East Main Street
Westminster, Maryland 21157-5026

Other books by the author:

The World of William Spry, Esquire: A Journey through Everyday Georgian England by a Recreated Eighteenth-Century Man

International Standard Book Number: 978-0-7884-3707-6

To the Last Voyage ; from which Nobody *ever* Returns

"To Die will be an awfully big Adventure ..."
A quote by Peter Pan (1904 - Eternity)

The date on this tombstone is 1702 ; one of many 18th Century tombstones in England which through showing the 'skull and crossbones' from Revelations in The Holy Bible and subsequently gave rise to the mistaken popular belief that a pirate was buried there ...

CONTENTS

Introduction

The author as an Early 18th Century Mariner aboard a full-size replica of HM *Bark Endeavour*, the ship commanded by James Cook which achieved an astounding 'voyage of discovery' 1768 - 1771. A tiny piece of the original ship went to the Moon with Apollo 15 and another in the form of a nail was carried aboard Space Shuttle Endeavour on it's maiden flight in 1971 on the bicentenary of Bark Endeavour's Return as an acknowledgement of past exploration in an age where *"Boldly Go Where Few have Gone Before !"* involved much the same planning, risk and reward.

"I must go down to the seas again, to the lonely sea and the sky ;
And all I ask is a tall ship, and a star to steer her by ..."

John Masefield (1878-1967) in his poem *Sea Fever*

Many readers of this book may already be familiar with the two lines above from the poem, but fewer may actually understand their meaning beyond the obvious romantic appeal. An earlier proverb grimly described the members of the human race as all belonging to one of three parts : the Living, the Dead - or Sailors. Compare this grim proverb to a quote by Doctor Samuel Johnson circa 1750 : *"No man will be a sailor who has contrivance enough to get himself into jail ; for being in a ship <u>is</u> being in jail, with the chance of* *better food and commonly better company."* The well-known harshness of conditions inside English jails in the mid-18th Century hints that a life at sea must indeed have been considered by most as something to be avoided at all costs. Britain being made up of islands meant such a life couldn't be avoided by everybody and a variety of sea-faring myths, traditions, adventures, romances, tragedies and victories by Doctor Johnson's time was already a firm part of the British heritage.

By the end of the first quarter of the 19th Century, reliable instruments and charts were widely available and carried on almost all ships by masters or navigators ; 'exploring voyages' by all nations to expand trade in the early part of the previous century had led to a few disputes when these ships met but also a frequent and frank exchange of information - in some cases, even between map-makers and scientists of enemy countries during wartime - and by 1815 most of the world with the exception of the unreachable poles had been more or less charted and mapped but leaving aside foul weather, shipwrecks due to running aground at night on shoals or reefs through indifferent or plain bad navigation still occurred. With the introduction of the quadrant, published astronomical data and compendiums, the accurate charting of magnetic compass variation and the availability of reliable mechanical time-keepers to use in accurately calculating longitude meant navigation required far less 'guesswork' than in the early to middle part of the 18th century but even as late as 1898 when all the above were available, a very experienced seaman named Joshua Slocum in returning from the first solo-circumnavigation of the globe under sail states in his book *Sailing Alone Around the World* that at that time maritime navigators acknowledged that a daily error of five miles in positioning was not uncommon at that time and even an expert in calculating using a sextant by the prevalent *'lunars method'* considered he had done very well in attaining accuracy to within eight miles of his true position ...

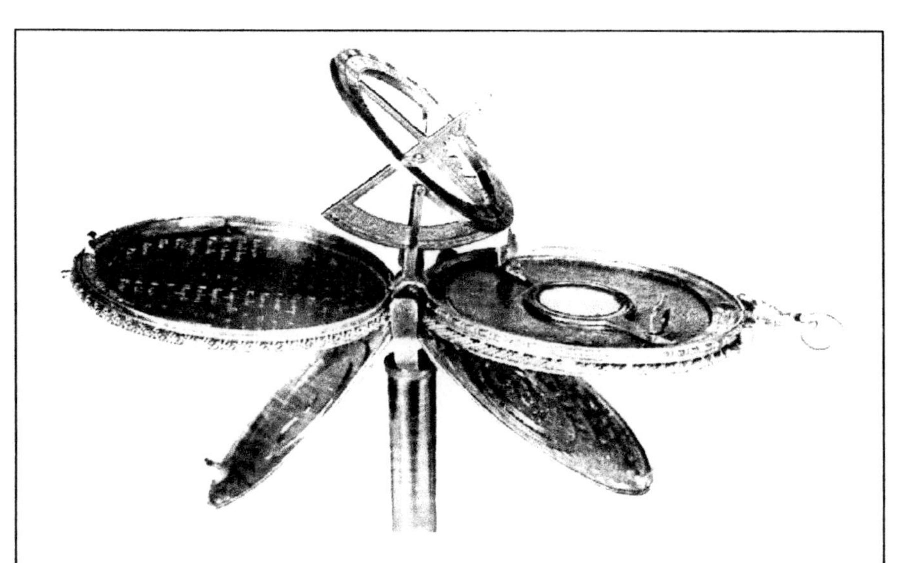

"Drake's Dial" : a unique piece of 16th Century maritime technology, made and inscribed in 1569 by Humphrey Cole, by tradition for the pocket of Sir Francis Drake. Made from gilded brass, the case is slightly oval, being three and half inches wide and two and a quarter inches deep. It contains a compass, an Equatorial sundial, the latitude of several ports, a table for calculating high-tide, a device which enables the owner to keep track of the sun's progress through the Zodiac, changing moon phases, compass-bearings of lunars at various ports, a circumferentor for use in coastal survey and a 'perpetual calendar'. Like the maritime chronometer when it was first introduced, such instruments were far beyond what a period navigator could afford but the technology involved was developed later into instruments which became available to all. *(Copyright National Maritime Museum, London)*

How did mariners aboard ships in the late 17th and early years of the 18th Century before the introduction of 19th Century navigational technology get to where they wanted to be? A captain or master of a ship that 'got lost' and ran out of water or food on a voyage risked the loss of the ship - and his life - at the hands of a mutinous crew ; but without the knowledge of balanced Time and Distance in a period where any and all available space aboard a ship was at a premium and the threat of scurvy and mutiny lurked just below the horizon, how did a ship captain know where he actually was at sea <u>and</u> how much food and water he needed to carry aboard his ship to feed his crew on any prospective voyage ?

A typical 18th Century maritime scene of a merchant ship loading cargo in a busy English port. Like many other vessels of her kind, the original vessel successfully sailed hundreds of thousands of miles during a period of 25 years.

Though the concept appears to have been known to Aristotle and Pythagorus, in the 16th Century Nicolas Copernicus established that the Earth moved around the sun rather than vice versa. In the year 1700 the superstition persisted amongst seamen for many years to come that the world was flat rather than round and usually manifested itself in the early 18th Century period in grumblings amongst a crew when voyages into unknown waters were proposed, especially into *The Great South Sea.* Although the former *'Sea of Darkness'* - the Atlantic Ocean - had been regularly crossed for many years, the Pacific Ocean still hadn't been accurately charted and large areas remained virtually unknown, becoming the source of some amazing speculations, claims and fantasies concerning sea-monsters, odd plants, weird animals and fierce cannibals who killed and then ate any trespassers. Superstition amongst seamen is well known - the crew of a buccaneer ship *'rounding the Horn'* on St Valentine's Day in February 1684 were forced further south than any other ship had ever been recorded as sailing by a fierce storm which lasted two weeks ; the crew blamed the captain for this very bad weather by his chosen topic of conversation on that day *'for discoursing upon women at sea being very unlucky, he had occasioned the storm'.* Such a belief today with the advantage of hindsight will be thought ludicrous, but the offer in 1714 by the British Parliament of a phenomenal £20,000 reward - the equivalent in modern values of over £10 million - for a practical system of accurately calculating longitude at sea did result in some equally ludicrous but well-intentioned impractical proposals. One of these proposals was to anchor ships exactly one half-degree apart all the way across the Atlantic in order for these to repeat a noon-time gun fired in London and another involved a surgeon who claimed that he had caused a patient to 'jump' though at the time the surgeon and patient were separated by a great distance and so suggested the application of his patent *'powder of sympathy'* ; the principle of this proposal being that when a person in London heard his pendulum clock strike Noon, he would dip the bandage from a previously deliberately wounded dog into a bucket of water containing the powder causing the same dog - then aboard a ship at sea - to bark at the same time, notwithstanding the distance between them. The first proposal was placed before The Board of Longitude but the latter evidently wasn't considered as one of their members objected on humanitarian grounds to the regular injuring of dogs. At that time it was commonly believed that Creation began at exactly 9.00am on Monday 23rd October in the year 4004BC ; a giraffe was the offspring of a mating between a leopard and a camel ; and a proposal had recently been made to explain the absence of swallows in wintertime was that they flew to the moon to hibernate - but the argument before The Royal Society was that swallows did no such thing and did indeed sleep through winter beneath the ice of frozen ponds!

Even today any sea voyage entails a degree of risk due to the enduring hazard of navigational error and although ship losses for any reason are rare, some still do occur on long-term voyages far from land and especially crossing areas subject to powerful seasonal storms and unpredictable weather such as the North Sea and the East China Sea - and in that place, even the threat of piracy still lurks. The successful application of the methods of navigation by the master or navigator aboard an early 18th Century ship from using his imagination, experience and art and in transferring this via a selection of mysterious instruments onto paper to create a vision of how to successfully and safely transport the ship to faraway places and return led to him being termed an *'Artist'* by uninitiated crewmen having little or no understanding of the overall concept. This book presents what was described during the past historical period in question as 'artistry' in an understandable, educational and entertaining form and answer the questions I posed myself over fourteen years ago when standing on the deck of the square-rigged pirate ship seen in this book. Some readers may question my references to buccaneers and pirates, but these men are good examples to choose for study in this respect as from the mid-16th to early 18th centuries several of these boldly performed some astounding feats of navigation, some of them voyaging deep into uncharted oceans - even if they did have questionable motives for doing so - and developing the long-haul voyages of the latter part of the 17th Century and at the end of the first quarter of the 18th Century which became known as *'The Pirate Round'*.
Protestant European nations and French Huguenot 'freebooters' did not choose to recognise the claim of Catholic Spain backed up by the Pope (who was also a Spaniard) after the discovery by Columbus in 1492 that the 'New World' belonged solely to Spain. Privateers, then

buccaneers - like Sir Henry Morgan - and finally pirates began robbing Spain of their 'New World' treasure and trade in the Caribbean and Pacific, which led to colonial settlement there by England, Holland and France. This meant that by 1680 many of the old Spanish coastal ports such as Cartagena, Port Bello, Vera Cruz and Panama had become no longer accessible by seaborne raiders through improved communications and defences such as patrolling warships and formidable fortifications equipped with powerful long-range artillery. The regular sailings of the Spanish *flota* treasure ships had dwindled to the unscheduled and infrequent sailings of smaller elusive ships and the once-yearly convoy was usually aboard a powerful warship. Growing commercial interests in the established European colonies in the Caribbean saw ship cargoes more commonly consisting of agricultural produce, timber, sugar, salt fish and ironware and these old and established cruising waters for privateers and pirates no longer yielded much portable profit through wealthy pickings as ships carrying gold and silver bullion were few and far between. Though the wealth carried by the yearly Spanish 'Acapulco to Manila Galleon' remained a temptation for years to come, taking this great prize with a single ship meant a major risk in 'running the gauntlet' of a voyage around the unfriendly coast of South America via Cape Horn and going well into uncharted enemy waters to wait for as long as it took for the galleon to finally arrive. By 1680, far-sighted privateer and pirate captains had already begun to look elsewhere for a source of easier loot ...
As a result, by 1690 rumours of even greater plunder began to circulate. The wonders of the Orient - gold, silver, diamonds, rubies, emeralds, pearls, spices, woven brocades and spun silks in addition to advances in science and medicine - were well-known before the time of Marco Polo in 1290 through previous Venetian and Genoese explorations sailing the 'inland sea' - the Mediterranean - and also establishing the 'silk road' by crossing into Asia at Constantinople to contact the great Moghul Empire in Persia and India as far as the Red Sea and the Arabian Sea. In 1599, England had established relations with The Great Moghul Emperor himself to seek trade with India which led to the founding of The East India Company in early 1601, their first trading post being established at Surat with a 'factory' at Bombay founded in 1662. The first major trading

fleet of the Dutch East India Company - the VOC - returned from The East Indies to Amsterdam in 1602 and the French soon after founded their own East India Company. The disadvantage in this was that to reach the Orient by sea by way of Africa, the voyage took a very long time - in the Atlantic, thousands of miles of dangerous waters had to be navigated to reach far enough south to be able to round The Cape of Good Hope before sailing much the same distance again to the north. Scurvy as a result of being unable to carry fresh stores on such a long voyage and unknown tropical fevers from 'touching' at Whydah or Benin decimated the crews of some ships to the point that these ships were hardly able to be handled. It was known through these trading voyages that many thousands of Muslim pilgrims each year descended on Mecca carrying offerings in the form of great wealth, many using the special seasonal pilgrim ships leaving from ports in India to Red Sea ports such as Jeddah and Mocha. Local pirates based in and around the Arabian Sea had often attacked these pilgrim ships ; by 1690, wars and local disputes with these pirates had seriously weakened the capability of the Moghul fleet to defend both their traders and the pilgrim ships. Like the British *'Triangular Trade'* described later, both meant sailing a ship along an awful long and risky route to get there but the pirate voyages would come to link the old roving waters of the Caribbean and the New England colonies with new seas and a new pirate base at Madagascar and other islands off the coast of Mozambique in east Africa.

In 1692, a privateer ship named *Amity* sailed from Rhode Island with a commission from the Governor of Bermuda to attack and reduce a French colonial base and trading post on the coast of Guinea ; but once at sea, the captain of this ship - Thomas Tew - made a speech to his picked sixty-man crew suggesting they become rich men instead by turning pirate. This pirate ship sailed across the Atlantic to Africa and rounded the Cape of Good Hope ; in cruising the Red Sea in July 1693 they attacked and captured the flagship of the Great Moghul *en route* from Surat to Jeddah and found aboard a *'kings ransom'* in plunder. When *Amity* returned from this voyage and docked at Newport in Rhode Island, rumours of each of the crew members enjoying a single share of £3000 meant that these seaman instead of earning an average of £12 to £30 *per annum* in wages had pocketed the equivalent of over a hundred years wages in just nine months. Though seamen knew the voyage from New England to Madagascar was long and tedious, with the potential threat of tropical diseases causing terrible losses amongst a crew and enemy warships in the Arabian Sea waiting to attack them, the commanders of privately financed proposed voyages of piracy fitted out in ports such as New York, Philadelphia, Boston and Newport could afford to turn away volunteer *'Roundsmen'* - who were hoping to sign on as crew eager to acquire such wealth - in favour of only the most experienced and *'seaworthy artists'*, and good navigators were especially required as the long voyage from New England also included crossing the Atlantic before passing into the largely unknown waters off the east coast of Africa.

The gaff-rigged Sloop ; a versatile vessel suitable for most purposes and especially when made from Bermuda cedar, a firm favourite with Early 18th Century pirates such as 'Blackbeard' operating in the Caribbean Sea. The sloop weighs around 100 tons, the hull is just over 50 feet long and shallow-draft which enabled the vessel to slip easily into shallow-water coves and most harbours. Capable of carrying a large area of sail-canvas, the sloop is very fast and manoeuvrable in most wind conditions but was unsuitable to carry large cannons because of limited deck-space. The sloop also had a limited cargo capacity making it unsuitable for carrying provisions for a large crew over a long voyage such as *'The Pirate Round'*.

In 1694 after a career spanning ten years, the most successful 'arch-pirate' of them all, Henry Avery - sometimes written Every - sailed a stolen vessel from the port of Corunna in northern Spain around Africa and in forming a small pirate fleet by linking up with Thomas Tew (then on his second voyage) and three other colonial-based pirate ships, Avery captured two large treasure ships carrying plunder at today's value of more than twenty-five million pounds sterling and causing rioting Muslims and Hindus in Surat and Agra in retaliation for their losses to imprison all 'foreign' traders - especially those of the East India Company - and caused a major uproar in Europe as reimbursement for the loss and future protection by warships was demanded by The Great Moghul himself. Tew was killed by a round-shot during one the above captures and Avery returned to England and subsequently disappeared, but after Captain William Kidd's failed 'pirate hunting' exploit in 1698 for the next four years - though Indian and Arab treasure ships still regularly sailed - the Indian Ocean became *'too warm'* for pirate ships. With the outbreak of war in 1702 between England and Holland on one side against France allied with Spain on the other, the prospect of further piracy suddenly faded as *'Roundsmen'* were actively recruited as privateers by both sides : any experienced navigator who could bring privateer fleets within reach of the 'Spanish treasure galleons' sailing from Lima to Panama or from Manila to Acapulco in The Great South Sea was keenly sought out by a English privateer captain. An English privateer venture there in 1704 had limited success, but another venture in 1708 did capture one of the fabulous 'Manila galleons'. It happened that the quickest and relatively easiest way home for these English privateers - in avoiding Spanish warships sent out to catch them after the alarm was raised and the prospect of sailing south against both adverse wind and current with the awesome prospect of going back around Cape Horn in winter - was follow in the wake of Sir Francis Drake and sail right across the Pacific between latitude O degrees at the Equator and 10 degrees North to reach friendly Dutch colonies around Java, before sailing on via the Dutch colony at The Cape of Good Hope and ending - literally - their 'round' trip in a complete circumnavigation of the globe. When the war finally ended in 1718 and the privateers found themselves out of work, many turned to piracy. After the elimination of the Caribbean pirate base at Nassau on New Providence Island, merchants and traders once again saw and experienced numbers of a new generation of pirate ships sailing on the traditional *'Pirate Round'*.

The author aboard an Early 18th Century pirate ship. Note the remains of the 'forecastle', the square-sail set above the bowsprit, the mizzen tri-sail yard extending beyond the mast - and a tiller-bar as in the year 1700 the wheel was a newly-introduced concept. This ship would be considered old-fashioned by 1725 but it was vessels such as this that were engaged in *'The Pirate Round'*.

The bold *'Pirate Round'* voyages existed roughly from 1680 until 1720 with a few very determined pirates still sailing the route after 1722 as due to their threat to a very valuable trade, French ships were then guarding the Persian Gulf, Dutch ships patrolled the Red Sea and British warships cruised off Africa and India all with mission to seek out and destroy pirates who had been labelled by all governments as *'enemies of Mankind'*. That these avaricious mariners achieved all this is not in doubt - but how did they find their way to all these faraway places ? As we cast off our moorings and set sail on our 'voyage of discovery', *Part One* of this work explains the situation before and during the historical period in question, *Part Two* offers a introductory guide to 'living history' and prospective 'do-it-yourself', *Part Three* gives more detailed information on the historical background, some useful figures and a look at later developments in the 18th Century that began to enable navigation to steadily become more a science than an art ...

THE PROBLEM

After the ship casts off moorings or raises the anchor to set sail with the wind or tide, as the local pilot leaves the ship to return by boat to his base so the work of the navigator begins and will continue until the ship reaches the approaches of another port and takes on another local harbour pilot or drops anchor perhaps thousands of miles away. The task before such a person entailed providing and collating information to sail the ship as fast as possible to where destiny lay and returning the vessel safely home. During *The Golden Age of Piracy* - a general period roughly from 1690 to 1725 - none of the 'traditional' navigational aids that viewers sometimes glimpse very briefly on modern maritime feature film or television drama entertainments such as *Hornblower* or *Master and Commander* were invented and all lawful and unlawful mariners employed the methods they had been taught or adopted similar stratagems to find their way to and from ports or their chosen cruising stations.

An early 18th Century trader bound for Africa leaves via the English Channel in the prevailing wind *'close-hauled on the starboard tack'* to clear Cape Ushant. The masthead pennant indicates both the wind strength and direction to the Master and the helmsman on deck.

In 1584, a Dutch mariner and cartographer named Lucas Janszoon Waghenaer drew up the first compilation of navigational charts, bound into a single book for use by Dutch mariners. This concept was thought so innovative by English mariners that they produced similar books and named them 'waggoners'. Spanish mariners and cartographers produced such a compilation containing charts and associated navigational information about the west coast of South America but the book remained a top secret until July 1681 when a buccaneer named Bartholomew Sharp captured the *Santa Rosario,* a Spanish ship bound for Panama and found a copy of the atlas aboard the prize. Sharp was later to write *"In this prize I took a Spanish manuscript of a prodigious value. It describes all the ports, roads, harbours, bays, sands, rocks and rising of the land and instructions how to work the ship into any port or harbour. They were going to throw it overboard but by good luck I saved it -and the Spaniards cried out when I got the book."* Sharp returned by way of Cape Horn to the Caribbean but on arriving at Jamaica was arrested for piracy - by the Governor, Sir Henry Morgan, an ex-buccaneer himself - and duly sent for trial in London. In October 1682, Sharp - knowing the value of the atlas - presented Charles II with a bound volume of the captured maps, inscribed on the front cover *'Dedicated and Presented by His ever loyal subject Barthw. Sharpe'.* He was - as a result - released from arrest and using a financial reward for the gift and a commission in the Royal Navy was able to return to the Caribbean to resume his career as a pirate *(Sharp died penniless and sick in a Danish prison on the island of St Thomas in 1698).* When this volume was examined, the importance of the maps had been instantly recognised by the Admiralty - at the time, the Spanish Ambassador in London pressed for Sharp to be hanged for piracy in desperately trying to regain possession of the captured atlas but was told to mind his own business and the book was immediately returned under armed guard to the London cartographer named William Hack for further translation and duplication. Incredible though it may seem to the reader - though it was not unusual for ships at this time to be carrying charts over 50 years old - an updated chart drawn in London in 1711 based on the maps within the captured Spanish atlas was placed aboard English ships sailing for The Great South Sea in 1739 ...

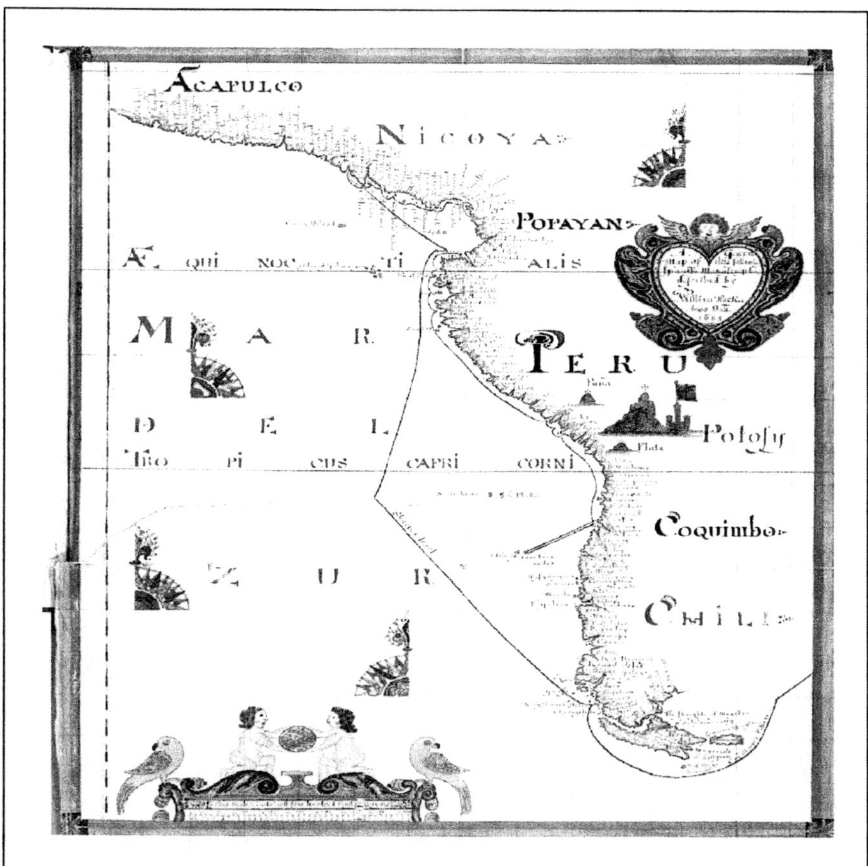

One of the hand-coloured pages from the Spanish atlas captured by Bartholomew Sharp in 1681. Such detailed navigational information falling into the hands of an enemy - especially England - had been dreaded by Spain for almost a hundred years. The route around Cape Horn taken by Sharp and shown here was only judged safe in the 'summer' season ; the equally notorious Le Maire Strait and Straits of Magellan 'short-cut' required both very careful navigation, watchfulness and luck as one ship still found out in the 1880's. Note that a latitude scale is shown but there is no longitude scale, hence the inaccurate shape of the continent of South America. The scale shown gives statute miles in terms of Leagues (see later). During this time-period, the Pacific was termed 'The South Sea' and the Caribbean as 'The North Sea'. ***(Copyright National Maritime Museum, London)***

Coastal navigation by local traders and fishermen in the early 18th Century still relied mainly on daylight and the naked eye using only experience, the lead-line and the anchor *(see later)* but if crossing deep water between continents, as various methods of calculating latitude were known from the medieval era, it was held by most navigators that the easiest way of finding a destination through an ocean-going voyage was not to sail out of sight of land and to sail north or south into a sea until the ship reached the known or anticipated latitude of the desired destination then to simply sail west or east *'running a latitude'* along the parallel checking this course daily and keeping a good look-out for as long as it took to reach where you wanted to be. A hazard here was that navigating this way - sailing on two sides of a right-angled triangle - meant these voyages imposed relatively safe but severe limitations on themselves regarding distance and sailed circuitous routes compared with the same voyages at the later end of the century as they navigated without using the advantage of trade winds and provisioning stops at mid-way stations ; adverse weather could cause some additional delay and a ship might then run out of food and water, but the ever-present hazard of shipwreck caused by the necessity of staying close to a 'lee shore' would be a constant worry. A buccaneer ship found the Galapagos Islands this way in late 1684 by sailing along two sides of a triangle (north along a coast to reach the desired latitude, and then steering west along it) as although they knew that these islands lay on the Equator, they did not know the longitude of their destination so could not plot a straight-line course. An additional hazard was that many destinations given in terms of latitude in the first half of the 18th Century included a varying degree of 'guesswork' by the mariners of the previous century. Even if you had a map, in using a chart one had to be aware than one nations understanding of the known globe was not necessarily the same as that of their neighbour ; in the year 1700 many English, Dutch, French and Spanish charts all varied in different amounts of latitude with longitude only shown at a given spot if someone had previously been there and had their calculations later confirmed by someone else, a mapping and charting process often taking many years and giving rise to the strange and changing shapes of some continents shown on early charts when compared with later versions. As an example, when France was accurately re-surveyed using these methods in the 17th Century, the King of France upon seeing the new maps claimed he had lost more of his territory to his state astronomers than to his enemies. This is why latitudes of seaports or islands were often accompanied in the late 17th Century by a simple reference as to how many 'days sailing' it would take to reach it from a given point. In another example, a voyage in 1686 by a buccaneer ship from the Galapagos Islands aiming for Guam - an island over three-quarters of the Pacific Ocean away to the west, given alternately in *'old English pilot books'* based on voyages in the preceding century by 16th English privateers such as Drake and Cavendish, as around 5500 miles away or around fifty days sailing - but in *'new Spanish pilot books'*, Guam was placed between 6900 and 7200 miles away, then over sixty days sailing. The buccaneer ship was already sailing on short rations due to both supply and storage problems and reached Guam after 57 days sailing, narrowly avoiding a disaster as they had just three days of these short rations left with the surviving crew in a very sad and mutinous state of sickness and starvation and who would have been destroyed even after landfall had not *'good fortune favoured them'*. After calculating his latitude at this destination, though later found to be accurate it was discovered by the same navigator that his previous calculations of position on the voyage to Guam were all incorrect mainly as he had not allowed for accumulated daily error through compass magnetic variation and the ship had reached Guam by luck rather than judgement. A later voyage in the Pacific, this time in the 1740's by ships of the Royal Navy using the same navigational criteria got into a scrape in being not being able to calculate longitude and sailing back and forth along a mistaken latitude in accepting a previous navigators' directions, 'missed' the Juan Fernandez Islands lying three hundred and sixty nautical miles off the then unfriendly coast of Chile, a voyage initially intended to cover a maximum of only one hundred and thirty-five nautical miles - perhaps three days sailing even in the adverse winds and current - but at the finish had taken over a fortnight as the latitude of the islands was found later not to be 33 degrees 30 minutes South as previously given, but 34 degrees 47 minutes South and over two hundred miles further to the west than anticipated. This same voyage had already had experience of miscalculated navigation after taking five weeks to round Cape Horn due to strong currents and headwinds due to a succession of terrible seasonal storms and in just one day - through sighting an unexpectedly dangerous shore - the flagship had to adjust its position from 87 degrees 57 minutes West to 78 degrees 28 minutes West, an distance of just under three hundred nautical miles. The flagship later attempted a similar voyage to Guam as the voyage described above by accessing the same trade winds ; due to the poor condition of both ship and crew they finally made it across the Pacific to an island north of Guam three months later, but this took almost twice their anticipated voyage time.

The Juan Fernandez Islands played a large part in exploration and exploitation of The Great South Sea in the late 17th- early 18th Century. They were discovered in 1574 'by accident' by a Spanish navigator named Don Juan Fernandez in seeking an alternative to the weeks spent fighting against adverse coastal currents and winds when sailing from the port of Calleo near Lima in Peru to Valparaiso in Chile ; he boldly tried a more offshore voyage further into the Pacific and succeeded in cutting the average voyage time of two months in half and in addition sighted the three islands. They became a valuable stopping-off place for buccaneer ships having rounded Cape Horn to take on food and water, before heading north towards Panama in search of Spanish booty. Alexander Selkirk - another navigator - was marooned on the larger island in late September 1704 by the captain of the buccaneer ship he was sailing aboard as Master and spent the next four years and four months there until being picked up in early February 1709 by a ship commanded by a privateer named Woodes Rogers - who went on to successfully attack one of the fabulously-rich Manila Galleons - who noted when Selkirk came aboard his ship that he still had his *'mathematical instrument and books'* and *'pieces that concerned navigation'*. Alexander Selkirk's situation contributed to the inspiration behind the first English novel, *The Strange and Surprising Adventures of Robinson Crusoe* written by Daniel Defoe and published in 1719. Selkirk himself died in December 1721 when First Mate aboard a Royal Navy ship sailing on an anti-pirate mission - the ship was damaged when it ran aground off Gambia and in sending men ashore to seek out materials for repairs someone brought back aboard a virulent fever which subsequently killed half the crew over the next three weeks.

The existence and approximate location of known pirates was well advertised whenever their presence was discovered by any colonial authority and pirates cruising near ports or harbours hoping to snare prey attracted patrolling warships or *guarda-costas.* Pirates had the added difficulty that they weren't welcome in port just anywhere and their intended victims usually sailed on a variation of a well-established route, so both survival and success for a pirate crew required a good knowledge of navigation and where their ship was during any voyage. Upon capturing a ship, a pirate's first joint-priority in searching for plunder was to question the master or navigator aboard that ship and if he proved to be more skilled than their own, then 'acquire' - by fair means or foul - both him and all his charts and instruments. One recorded example of why pirates did this happened in 1720 to pirate captain Bartholomew Roberts - by no means a bad navigator - in sailing east from the Lesser Antilles Islands in the Caribbean aiming for Brava, the southernmost island in the Cape Verde Islands group off the coast of West Africa ; through an error in navigation and experiencing adverse winds, he 'missed' the islands and arrived too far to the south - he could not sail the ship north against the prevailing trade winds and was forced to run back west for 2000 miles across the Atlantic with only fifty-four gallons of drinkable water remaining for his crew of 126, reaching the coast of Guiana in South America some weeks later. Only Roberts' well-established past reputation as a skilled captain saved him in this instance from *'marooning'* or cold-blooded murder by a sick, starving and very thirsty crew who had been reduced to a cupful of water each day - and when that ran out, drinking a mixture of seawater and their own urine. The reader might think that their experience put the crew off the idea, but they set out again the very next year and this time successfully made it to Africa - only to be caught at sea by the Royal Navy, with 'Black Bart' Roberts being killed in action and most of his crew captured and subsequently imprisoned or executed.

From 1690, charts and maps showing the course of seasonal prevailing trade winds - and of course the *'Doldrums'* , those areas where winds crossed or suddenly ceased leaving a ship dangerously becalmed for long periods - and the associated ocean currents around the globe slowly became available in print but steadily expanding trade and equally, wartime naval strategy still decreed an overwhelming need for even more accurate maps, charts and instruments and above all a simple, reliable and practical method of accurately navigating and charting. From before that year the weight of the scientific minds of all European nations had increasingly turned to solving the problem of discovering a far more reliable practical method of navigation at sea ...

An 18th Century pirate ship attacks an armed French East Indiaman off Madagascar by *'crossing the stern'* and taking the wind gauge. Death and damage notwithstanding, navigation must continue at the conclusion of any naval engagement.

THE THEORY

Basic Seamanship

Though ship-handling was the Masters responsibility rather than the Navigator, these two posts were often held by one and the same person and the following would directly affect setting and the maintaining of a ships course toward any destination.

The author goes aloft aboard ship using the mizzen-shrouds. When compared with their equivalent 'landsmen' workers ashore, a surprising number of Able Seamen appear to have been able to read and write in the early part 18th Century which was probably due to the desirability and advantages in 'rating' by grasping the concept of basic navigation. There was an incentive for this aboard a ship as sudden losses and illness could and did regularly occur and any able seaman might be called upon to take on the duties of another during such an emergency.

In the diagram below, use the outer ring for a vessel such as a gaff-rigged sloop and the inner ring for a square-rigged merchant ship. If the wind is blowing onto the bows as shown, forward speed would be greatly reduced and as neither ship can sail a course straight *'into the eye of the wind'* could not go *'closer to the wind'* by entering the area shown in black. As a wind blowing abeam changed direction (*'veered'* or *'backed'*) towards the bow, then the angle of the yards holding the sails on the masts of the ship would have to be adjusted by the crew - terms such as *'close-hauled'* apply to either beam of a square-rigged ship - and should the wind direction continue to move towards the bows, eventually the course of the ship would have to be changed and turned away from the *'eye of the wind'* to avoid the vessel being *'taken aback'* by holding a course which entered the area in black. For example, a square-rigged ship sailing the course indicated with the wind direction as shown would be *'close-hauled'* in area **B** and *'close-hauled into the eye'* in area **A** with the yards set at almost 45 degrees to the masts : if the wind direction moved five degrees either way from the direction shown, then by necessity so must the course of the ship.

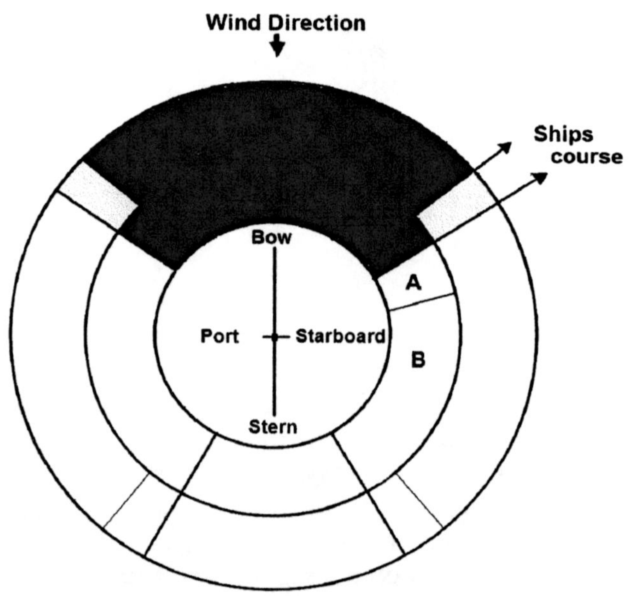

As the wind increased or decreased in strength, the speed of the vessel with the same sail-plan - especially within the grey areas - would also rise or fall ; in a steady wind, the ships speed (in knots) would fall in proportion to the wind direction in blowing towards the areas shown in white then moving direction towards the grey areas forward or astern. Note that when the wind is blowing towards the bows, if the wind strength fell to a couple of knots then a ship on the course shown will hardly make any forward progress at all. As previously stated, with a destination in mind the ship can only make forward progress *'in the eye of the wind'* by *'tacking'* alternately in a series of 'zig-zags' (to either port or starboard as *'close to the wind'* as possible, then vice-versa) but forward progress towards your destination in terms of nautical miles would be painfully slow and be closer to the destination by a small fraction of the nautical miles actually sailed by the ship during *'tacking'*. In a steady wind but a changing wind direction, in holding a set course but adjusting to set the best sail-plan a general guide follows :

The closer to the eye of the wind the ships course is, the slower the ships speed.

Wind abeam moving towards the stern, speed will gradually drop by 50%

Wind abeam moving towards the bow, speed will gradually drop by 80%.

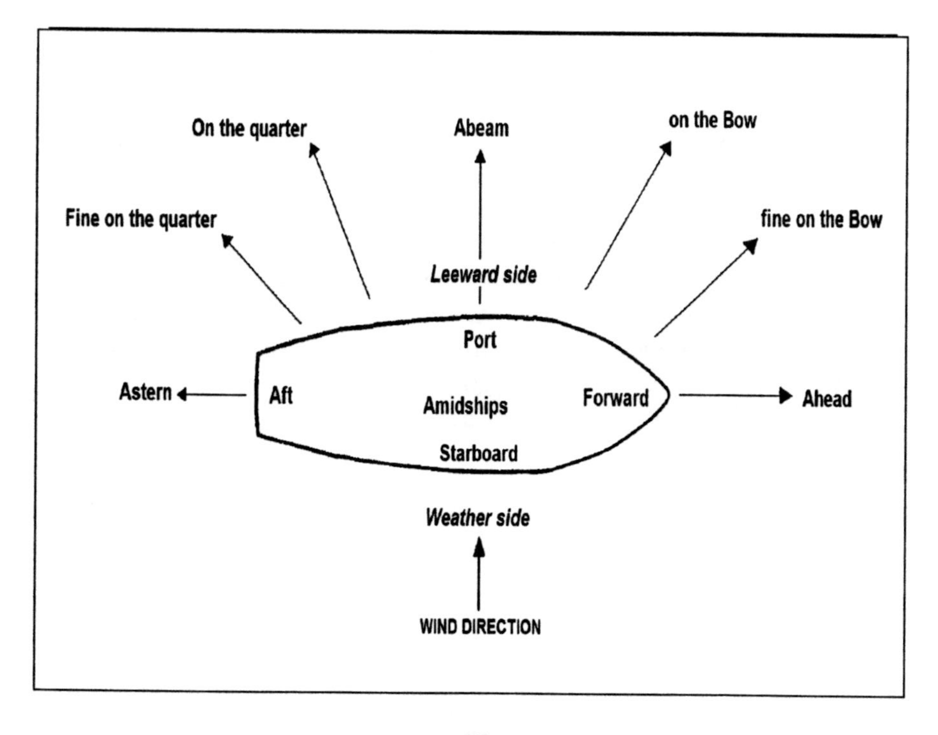

Each individual ship has its own particular sailing characteristics, but in general a square-rigged warship or merchant vessel would sail best when *'large'* (wind direction blowing abeam or a quarter) to *'running before a wind'* (with the wind direction blowing over the stern) but a 'lateen' or triangular-sailed vessel - such as a sloop or schooner - would sail best by ***'sailing by a wind'*** (wind on a quarter).

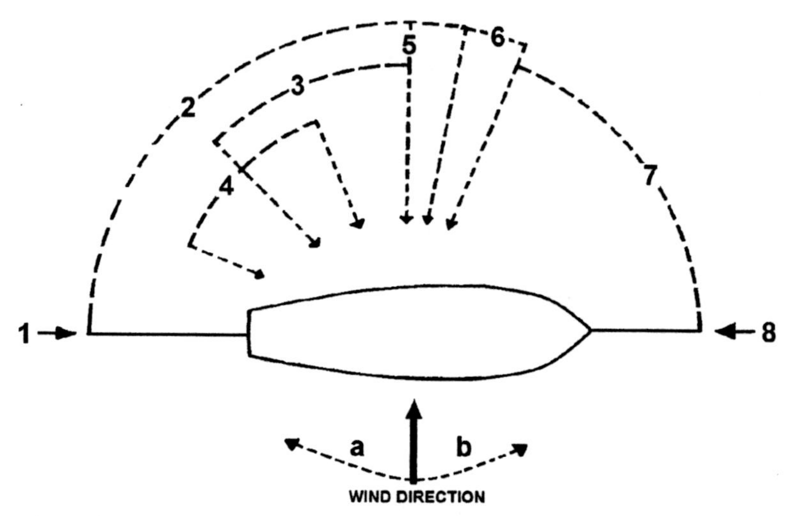

In the diagram above , a wind blowing towards the ship within the numbered sectors would be described by a mariner as follows : 1. 'following' wind 2. 'fair' wind 3. 'leading' wind 4. sailing *'large'* or a 'quarter wind' 5. a *"soldier's wind"* 6. 'scant' wind 7. 'foul' wind 8. *'the eye of the wind'* **a**. wind is *'veering'* (moving sun-wise) **b**. wind is *'backing'* (moving anti sun-wise). The term *"soldiers wind"* is not exactly complimentary as it was said that a steady wind from that direction meant the ship could be sailed and handled easily requiring little or no adjustment to the sail-plan.

A square-rigged ship under topsails and the fore main-course before a *'fresh'* following wind.

Prevailing 'Trade' Winds and Currents in the Atlantic Ocean

In the following diagram, prevailing trade winds are shown as broad arrows and strong currents as dotted lines - these were well-known to several individual navigators by 1680 but not generally known by all navigators. It can be seen how the *'Triangular Trade'* developed in the Atlantic as ships sailed 'sunwise' to reach destinations as a navigator would try never to plot a course by sailing against the prevailing wind or current. A British merchant ship in the year 1700 sailing to the Caribbean from England would first carry a cargo of cheap consumer goods to Africa, perhaps putting in at the Canary Islands ; on to the Cape Verde Islands to load salt to drop off in Benin, then from Africa his course would be due west across the ocean, before setting out for home via the coast of North America to drop off part of the original cargo in terms of glass, ironware and consumer goods and part of a cargo of sugar or tea and collect tobacco and rum to carry to Bristol or London. The original cargo for Africa could be weapons such as flintlock muskets and/or cheap mass-produced ironware and trade goods to be sold or traded for slaves to work on the sugar, tobacco and cotton plantations of the West Indies and the British Colonies in America ; then shipping those particular

British Colonies in America ; then shipping those particular products along with rum back to England on the final leg of a *'round trip'* which depending on the number of planned or unplanned stops took on average between six and twelve months to complete. The Leeward Islands of the Lesser Antilles in the Caribbean Sea via The Canary Islands or Madeira were the destination for Spanish ships from Cadiz heading for the Caribbean ; the Straits of Florida had been used by the Spanish *flota* to return to Spain from their ports and bases such as Portobello and Vera Cruz each year, but several disasters had occurred in bad weather on the notorious Bahama reefs south of Florida due to navigational difficulties, poor visibility and the lack of any 'sea-room' in which to manoeuvre. To avoid this passage, the famous *'Windward Passage'* between Cuba and Hispaniola started to be used by both galleons and traders - but this became a highly predictable route and after 1620 was well within reach of small groups of buccaneers operating from the western end of Hispaniola and even more so later when all these groups formed themselves into *'The Brethren of the Coast'* and became based on the nearby island of Tortuga. As French buccaneers had already settled on Hispaniola, after 1655 - when Jamaica was captured by the British - the 'Windward Passage' route was judged to be very risky for Spanish treasure ships, who then returned to the Florida route which by that time had become a focus for more French buccaneer operations through several established colonies west of Florida and from 1705 a nest of pirates had established themselves on New Providence Island *(modern Nassau)* in the Bahamas. British buccaneers operating from the old base of Cagway on Jamaica - renamed Port Royal after the restoration of Charles II in England in 1660 - then used the prevailing wind in the Caribbean to transport their men by ship to the Isthmus of Panama where they raided the Spanish coastal bases there such as Porto Bello and Puebla Nuevo, making a quick getaway on their ships afterwards in the opposite direction using the same wind then blowing from the starboard quarter. Buccaneers also landed on the isthmus from ships and crossed overland to attack Panama itself, either re-crossing the isthmus (considered to be very risky after an alarm had been raised, especially without native Indian guides) or simply steal a ship in the Bay of Panama to return with their plunder via Cape Horn. Popular trading bases in Africa for ships to aim for were Whydah and Benin on 'The Slave Coast'.

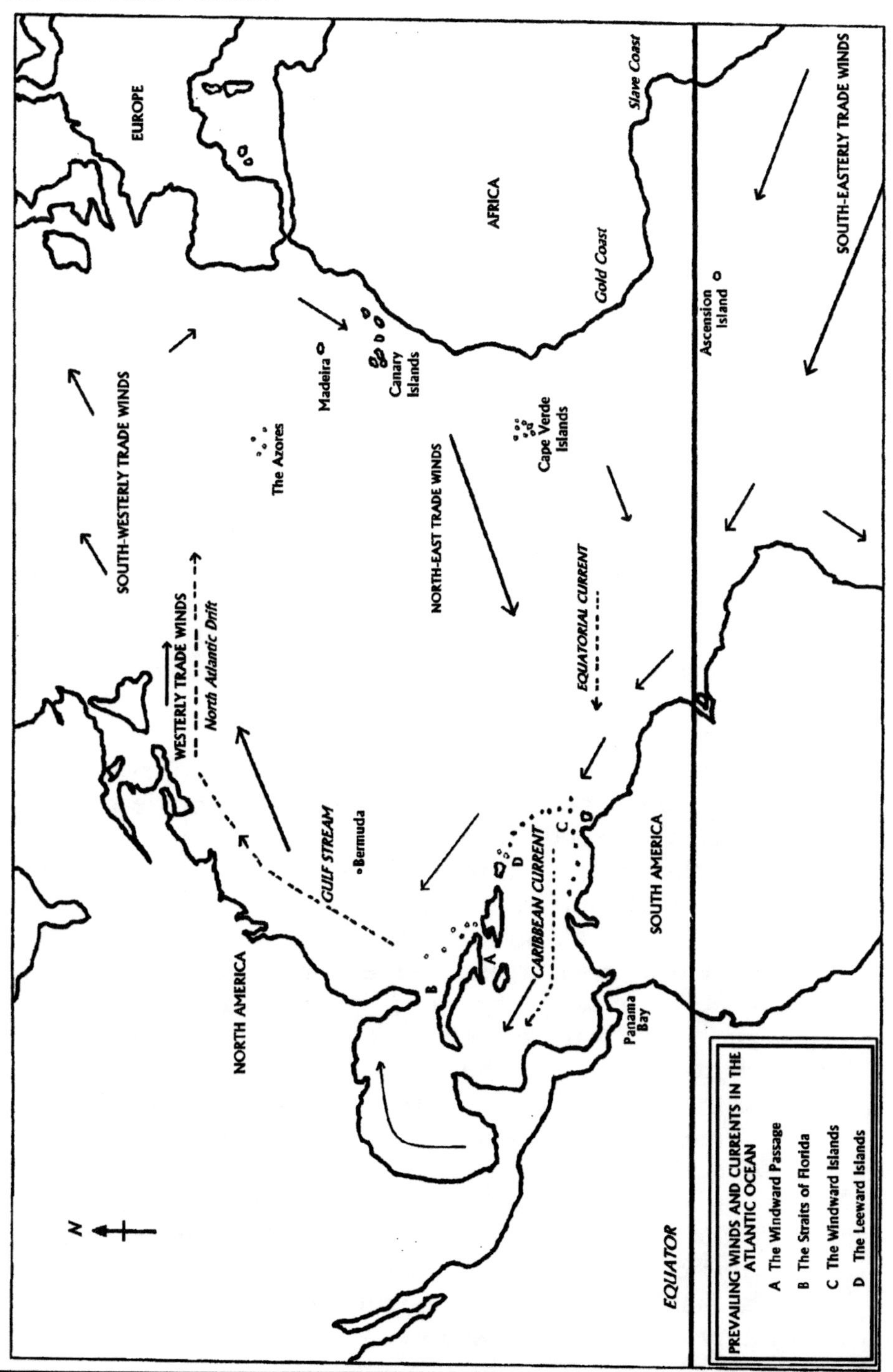

To sail from the Caribbean to Africa and take advantage of favourable currents and trade winds, a ship would set a northerly course before heading west and then turning southeast - but if you made an error in navigation crossing open sea and passed to the south of The Cape Verde Islands you would then be faced by strong 'headwinds' and an adverse current so not be able to reach the mainland without some serious delay and a lot of hard work for the crew in regularly 'tacking' to port and starboard ; this is what happened to pirate captain Bartholomew Roberts in 1720 who faced with that situation had to sail his ship all the way back to South America using the south-easterly trade winds and the Equatorial current, making landfall at Surinam literally just in time to save the ship and his starving crew from destruction. A similar pattern of prevailing winds and currents occur in the Pacific and Indian Oceans and these too were later charted on a changing seasonal basis for navigators to be able to use aboard ships trading with the Middle East, India and the Orient. A merchant ship sailing from London in 1704 on a trading voyage to India saw the ship away at sea for two years before a return to England carrying a cargo of pepper.

Basic 'Plain Sailing' Navigation

Navigators at the end of the 17th Century had inherited the experience and techniques of their maritime predecessors; some ancient methods through continuity of use were very well established and through this and the expense of *'suspicious new-fangled'* navigational instruments based on the same principles continued to be employed even after the introduction of instruments. One of the tried-and-trusted methods used in antiquity was to use an arm stretched out horizontally and with the hand define latitude : measuring the height of Polaris - the Pole Star or the North Star - above the horizon, the breadth of a single finger defined about 2 degrees, the thickness of a wrist about 8 degrees and by the full span of a hand about 18 degrees. Try it - sailing along the same latitude at night means keeping Polaris the same height or angle above the horizon as you hold your course so this simple 'rule of thumb' will work after a fashion - but you might take into account for continuity that the dimensions of the hand of a young man going to sea for the first time compared to the size of a hand possessed by a veteran seaman after many years of hauling on tackle would engender a variation. The obvious next step for ancient mariners concerned with navigation who understood the above principle was to develop a standard unit of measurement and incorporate this unit into an instrument - which is exactly what they did.

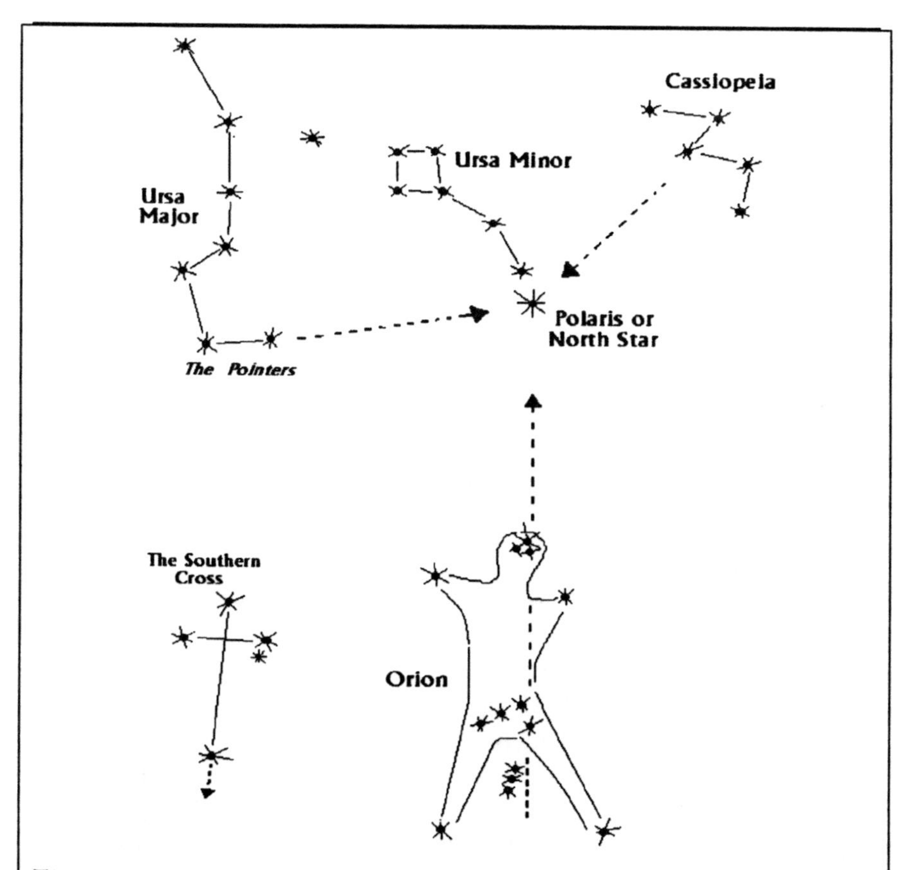

Though ancient mariners sometimes refer to star constellations by different names, these are just four constellations that were constantly used by them in navigation and easily recognisable today. The advantage of *Orion* is that this constellation is very distinctive and can be seen from both the northern and southern hemispheres - a theoretical line drawn through Orion (as shown) will eventually reach the North Star but mariners would use a straight staff to do this aboard ship. In the southern hemisphere as South isn't marked by a star, mariners in the 18th Century would come to use the five-star constellation known as *The Southern Cross* : twice the distance of the long axis of the cross in the direction of the dotted line marks a position roughly above the South Pole.

The author aboard a traditional vessel in the Mediterranean Sea was shown a recreated instrument known as a *kamal* to 'find' a latitude using Polaris - nothing more than a square piece of wood with a hole bored in it into which a length of knotted string was fitted and held in the teeth, but it worked in finding the latitude upon which your port of destination (but only if someone had been there previously) or port of departure lay ; the remaining decision would be whether to sail east or west on that latitude to reach either. In a similar fashion, every Boy Scout used to know that when the shadows are shortest the sun is at its highest in the sky and so could establish local time at 'Noon' to a few minutes. On a clear night as long as their ship was north of the Equator, mariners could see The Pole Star, which stays very close to True North ; experienced seamen could roughly count the passage of hours at night by glancing up at the changing positions of the stars in the constellation of *Ursa Major*, also named *The Great Bear, The Plough* or *'The Big Dipper'* : in the early 16th Century an instrument called the 'Nocturnal' was developed from this principle to give a time at night which would be accurate to fifteen minutes. As long as the ship was north of the *Tropic of Cancer*, the sun would always be due south at Noon and 15 degrees farther to the east for each hour nearer to sunrise and consequently 15 degrees farther west for each hour nearer to sunset. Local time could be reset each day at noon and kept track of during the rest of the day by using half-hour or one hour sandglasses for as long as the diligent person given the job stayed awake (or alive) and had not omitted to turn these glasses when they ran out of sand in very bad weather or if a round-shot from a pirate or warship had disrupted the usual system of time-keeping.

A mariner in the 18th Century used either or both of two methods of calculating a position at sea ; the accuracy of the first depended largely on accrued maritime experience and the second method on available instruments, printed documentation and having the mathematical know-how to be able to use them. Taking the first method the principle is that if you begin a sea-voyage at a known location and at a known time, by keeping a daily written record of time, speed and distance taken at intervals throughout a 24-hour period you can calculate a position at any given time at the end of the day ; if you know where you started from, by applying speed and compass heading noted at several times during the day you can chart your progress. Basically, setting sail at noon on a compass heading of due west for two days at an average speed of five knots would give you a figure of sailing 240 nautical miles over 48 hours in that compass direction ; this course you can then draw as a straight line on a chart (if you have one) including the regular checks and adjustments for speed or sail-plan and any change of course. If the voyage progresses out of sight of land, unfortunately this position can become more and more inaccurate due to variables which couldn't be accurately measured as the ship in open water <u>hasn't</u> actually sailed along the straight line drawn on the chart.

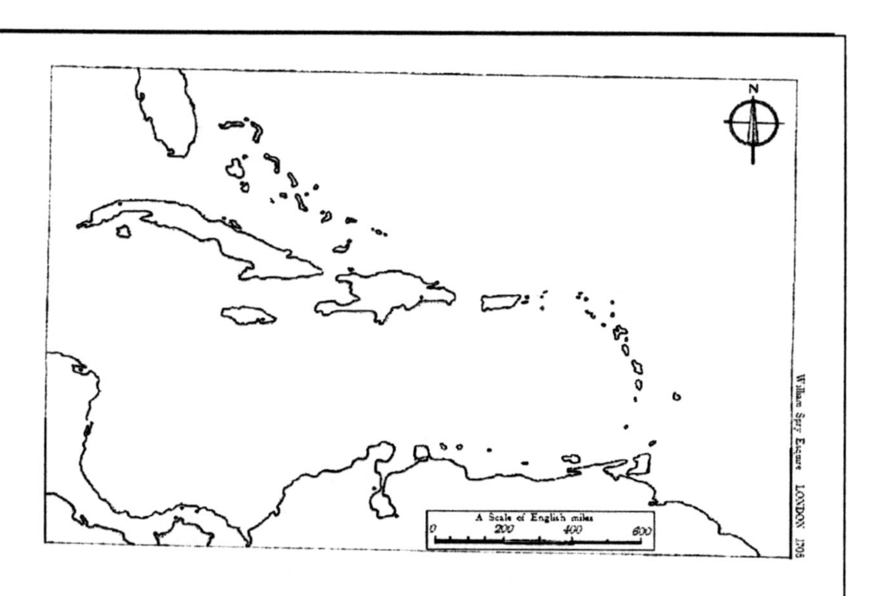

Drawn by the author using a pantograph, the first in a series of basic charts when enlarged to full size can be used in learning and practicing basic navigational techniques. This chart shows the Caribbean Sea, the Lesser and Greater Antilles, part of South America and the isthmus of Panama and was designed as the first step to explaining and understanding the difference between a map and a chart.

This method is known as *'Dead Reckoning'* (more correctly, *'Deduced Reckoning'*) and by applying past experience and charts - if the navigator possessed either - can also accommodate sightings of any known landmarks and adjust for the variations of current and wind to give a more accurate 'approximate' position. The second method used is more accurate - but it depended on the skill of the navigator in mathematics and a series of instruments which would only be developed and become gradually available over the span of almost the entire century. Based on the principle of measuring at a known time the height above the horizon of the sun, moon, a planet or a known star a navigator could - by a calculation or by using the tables in a periodic nautical almanack - compute a latitude (the position of the ship north-south). By calculating the time-difference of sunrise at the longitude of the port you departed from and the same event seen from aboard your ship at sea in local time, you can calculate longitude (the position of the ship east-west) - but though this principle was understood at the beginning of the century, a navigator at sea required an accurate time-piece aboard the ship to make this calculation easy and reliable and such time-pieces would not be available for maritime use until the very end of the 18th century ; and in reference to the lack of accurate charts, as the Astronomer Royal wrote to Samuel Pepys of the Navy Board in 1697 ... *"It is in vain to talk of the use of finding the Longitude at sea except you know the true longitude and latitude of the port for which you are designed."*

Draw your computed latitude and longitude by horizontal and vertical lines on an accurate chart and the place where the two lines cross is your position at the time of the observations made *(an example is shown in Part Three)*. This is known as the *'Observation'* method of navigation ; theoretically, by using observation a navigator does not have to consider the variables of speed, current or wind as an observation would show any miscalculation in his comparative *'dead reckoning'* but usually a ship would not strictly rely on either method alone and prefer to employ both methods performed by more than one person aboard and check different logbooks against each other, with each observer marking in their own log against their calculations a *DR* for the first method of finding position and *OB* for the second method and each log studied in total would provide a check against each other in terms of accuracy and offer a mean or 'average' position. Any one of these individual positions which showed the ship far from an 'average' *could*

indicate that someone had done their sums wrong - but many ships had only one navigator and as will be seen *could* also show that only one of the assembled navigators actually knew what he was doing and where he actually was. In the last quarter of the 17th and the very early 18th Century two whole fleets were lost at sea - by England off the Scilly Isles during daytime and France off the coast of Venezuela during night-time, both involving a great loss of life by ships being destroyed by sailing onto rocks or reefs in thinking they were safely in deep water. The tragedy in both these situations was that some seamen aboard these vessels *knew* where they actually were by accurate *'dead reckoning'* and the ships in the fleet were in great danger of running aground but their opinions - voiced on both occasions - were discounted by the admiral in charge as far from the 'average' of the fleet navigators. Another two obvious difficulties for navigators was that any accurate observation relied on reasonably clear weather to actually be able to *see* the coast or the celestial body selected and a reasonably steady deck from which to take an accurate reading.
The Navigator would also be subject to the other 'occupational' hazards of a long-term voyage by living in cramped conditions which were often bad and as a result of the ship-board diet could be affected by the scourge of the period seaman ; Scurvy. One merchant ship in the 1700's was stocked for a long voyage with a weekly allowance for the crew as follows :

Beer 7 gallons
Bread, hard biscuit or flour 7 pounds
Salt beef 3 pounds
Salt Pork 2 pounds
Dried peas 2 pints
Barley or oatmeal 3 pints
Sugar 1 pound
Cheese 4 pounds
Butter 1 pound
... and a gill of brandy or rum at the discretion of the Master

The Captain and Master - as was the usual practice aboard ship - packed away some separate stores for their own private consumption including a pig, a goat to give milk and several chickens to provide eggs which of course would necessitate provisions to feed and water these animals until they were themselves eaten. The above list is typical but doesn't mention fresh water - there would have been an allowance and provision of water kept in the casks in the hold for cooking purposes, though after a few weeks at sea this liquid would become the green-tinted sludge described by many seamen and replaced whenever possible, hence the allowance of beer. Outward-bound British ships often stopped off at one of the Cape Verde Islands off the west coast of Africa to take on fresh water and also salt, to use in trade and further provisioning. Though the seamen might also be able to catch fresh fish to augment this menu during the voyage, stored foodstuffs were all dried or salted and by modern standards were not healthy - note the lack of any fresh fruit or vegetables in the above list.

The 'living space' aboard ship below decks for the crew ; the boxes on deck are the sea-chests containing the personal belongings of the crew which serve as seats at the mess tables. All these fittings could be stowed away to make more cargo space but the navigator aboard ship would have the use of either a chart-room or the Stern cabin to lay out his charts and instruments and perform the necessary calculations without distractions. Through hatches and gratings on deck could be removed to admit light and air, the lower decks of ships were usually lit by lanterns or 'glims' making them permanently gloomy. The tangled rope seen on the table is the authors' log-line after use by visiting children !

Rats aboard ship were a pest that had to be kept firmly under control and as these vermin fed on ships' stores the crew usually had no objection to catching and eating rats in return - but period accounts of a mysterious shipboard illness linked to scurvy and likened to 'infectious jaundice' was probably *Leptospirosis* caused by the crew coming into contact with rats' urine as the harmful spores within this medium can survive for several weeks in the moist environment below deck and enter the human body by contact through the eyes, nose and mouth or any open wound. Though better off than their equivalent 'landsmen' and sea-faring conditions were a little better in the Royal Navy of this period, the journals of period merchant seamen note that the only unsalted meat they ever ate aboard ship came in the form of a ships' rat deliberately caught alive in a cage-trap. Although they and their shipmates also caught and cooked a variety of seabirds, fish, sharks, whales - and delicious turtles in the tropics - in general terms, after three or four weeks at sea the crew ate their food allowance in the knowledge they would starve to death without consuming it but trimmed or snuffed the candle in the mess lantern below decks so they didn't have to actually *see* what they were eating. One period mariner also notes how his shipmates never washed their bodies, rarely shaved using fresh water and seldom changed their shirts, how the fierce equatorial sunshine seared the decks in the daytime causing his shoes to stick to the melted tar in the deck seams, how the stench from the bilges was intolerable - often noxious to the point of being dangerous to the crew - and on a badly-run ship how the amount of bugs in the form of cockroaches, fleas and lice grew in abundance. It is both recorded and common knowledge that the captains and shipmasters of some ships were cruel tyrants in the treatment of their crew, especially regarding casual punishment ; if your captain turned out to be a combination of one of those and a 'skinflint', the entire voyage could turn into a crippling nightmare with very few happy moments which the navigator would also have to endure. There were certainly some unscrupulous captains and masters about but word quickly spread amongst seamen about their names and ships, with only really desperate men signing aboard these as crew.

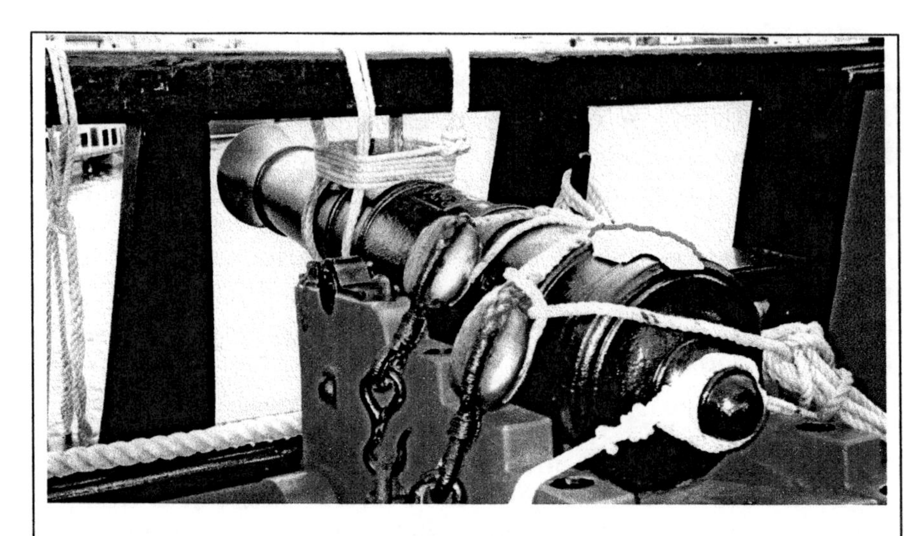

Warships carried many muzzle-loading guns of varying calibre, but merchant ships usually carried at least one cannon on deck to be used for signalling or saluting purposes - a gun could be fired as a directional warning in bad visibility, as a signal the ship was 'in distress' or as a time-keeping warning to the crew if ashore. The guns could be used in self-defence against belligerent native canoes or pirates but even though it was a common practice to keep the guns loaded, before use of the gun all the tackles seen would have to be 'cast off' as this small deck gun - a 6-pounder calibre carriage-gun - weighs almost a quarter of a ton and is securely lashed to prevent any movement in bad weather to avoid *'a loose cannon'*, the origin of the modern term.

Slave ships - sometimes named *'blackbirds'* - were never popular with ordinary seamen as the combined stench, sickness and disease brought on through horribly cramped and unsanitary conditions aboard resulted in many of the crew dying on the voyage along with large numbers of their human cargo. Some crews were 'hoodwinked' by captains who spread the word that their prospective voyage involved only 'honest' trading and only announced after putting to sea their destination was to be The Slave Coast - or even further, into The Great South Sea - by which time it was too late for the crew having signed-on for the voyage to complain or do anything about it except by risking trial and execution after a charge of mutiny.

The rank or 'rate' of Navigator aboard a ship depended on particular circumstances - he was sometimes ranked and named the *Pilot* and as a specific entity might be rated below the Master or in some cases, above him. But - a navigator would always be considered one of the ships' professionals and in many cases was the most literate, numerate and educated man aboard. A Master would train his Masters' Mate rather like an apprenticeship unless that person was already well-qualified ; navy officers were obliged to train in navigation from an early age and some specialists in this field - including potential Masters' Mates from Able Seamen who were able to read and write - later opted to attend a two-year course on maritime navigation at a shore-based college (instigated for the Royal Navy by Samuel Pepys in the 1660's) though shore-based education was widely acknowledged to be no substitute for practical experience gained at sea. A navigator would also have to have some experience of ship-handling and ship-building ; the gradual growth of weed and barnacles on a ships hull would slowly impede its sailing qualities and progress through the water, requiring a regular (between three and six months) maintenance through *'careening'* (from the latin word for keel) : the unshipping of all tackles, yards, guns and stores before either hauling the vessel out of the water, placing it in dry dock or simply using a convenient spot to tug it over onto one side then the other using heavy cables to enable the crew to chisel or burn off the attached marine growth then chisel off the old protective layer to restore the thick protective daub made up of substances such as tar, sulphur and tallow. At this time - especially if the vessel had sailed in warmer waters - an inspection of the lower hull, the keel and rudder would take place concerning the activity of *Teredo Navalis* (often described as 'shipworm' and is actually a species of clam) which in boring holes and tunnels in wooden planking below the water-line could seriously affect the integrity of a vessel if not dealt with. Ships' hulls often had a replaceable wooden 'sheathing' of extra timber to prevent this damage, and prevention of both 'shipworm' and marine growth developed into the *'copper-bottomed'* vessels of later decades. The interior of the hull could also be cleaned whilst empty by scrubbing it with vinegar. The *'trim'* of a ship in off-loading or loading cargo and also in using up water or ammunition would also affect its sailing qualities, and require the remaining stores in the hold to be moved to adjust and balance the trim ; an over-loaded cargo ship would be termed *'cranky'* and be in

danger - especially during bad weather or a sudden squall - of *'broaching to'*.
A ship often carried two sets of sails, one set of canvas for sailing in prevailing light weather and another set when expecting *'heavier'* weather ; being caught by a seasonal storm in the *'Roaring Forties'* and still having light canvas aloft meant that if something suddenly *'carried away'* it could result in serious damage and shipwreck. Some ships would sail *'closer to the wind'* than others, and doing so risked serious damage with any slackness or lack of attention at the helm but with any ship with the wind blowing from the bows the speed would be greatly reduced. To sail a course against a prevailing or adverse wind meant a time-consuming and exhausting manoeuvre known as *'tacking'* ; a series of alternate zig-zags to port or starboard and vice-versa which covered many sea-miles but on a chart meant slow progress towards your destination. Tide variance meant the crew of a pirate ship in 1684 found themselves in trouble after dropping anchor in the gulf of San Miguel in the Bay of Panama - though the rise and fall of the tide in their usual cruising station of the Caribbean Sea was almost imperceptible, the fall of tide over on the other side of the isthmus was suddenly discovered to be over twenty feet and at low water left their ship stranded on a mudbank. The same crew had endured a previous upset when their lookout sighted what he thought was 'white water' above a hidden reef obliging the helm to be put hard-over - but the suspected reef turned out to be a surface disturbance caused by a vast surface shoal of millions of tiny fish ...

A Master aboard a vessel, along with the boatswain and all other 'officers' would by necessity have to take all this and many other relative sea-faring aspects into account and the pressures on a navigator to successfully get the ship to a destination as soon as possible was immense. In having only a single navigator aboard, a ship relied solely on his knowledge, skill and personal experience to reach a destination - if he was lost for any reason, the ship was in great peril - but the reader may already have noted that even having an experienced navigator aboard, a ships position at sea in this particular historical period could only really be plotted within a given area : a position on average - depending on the past weather and the duration of the voyage - somewhere within a circle of between ten and several hundred miles radius, an area that could *in extremis* cover several hundred square miles. Any *rendezvous* between ships had to be fixed at a point or a port on land and only within a time-period of several weeks rather than a particular day ; a difficulty that could severely disrupted or even destroy a planned strategical move by a navy, especially when ferrying troops long-distance for a combined amphibious operation involving a necessary degree of surprise against a island or coastal base held by the enemy, as later occurred during the worldwide conflicts of 1690-1718, 1739-1748 and 1754-1763.

Shtandart ; a unique replica of a 1705 warship

Good training for a British-born navigator in the expanding trade boom of the late 17th to early 18th Century was said to be in the maritime *'fetching trade'* across the English Channel and North Sea, and in reading between the lines I take this to mean that though these voyages were relatively short-haul it was considered that if a navigator could handle these trips subject to unpredictable weather, lee shores, hidden rocks, shifting sands, an unexpected cannonball from a passing ship or shore battery through regularly changing foreign policies and negotiate passages into some difficult harbours or ports and river estuaries such as the Thames, Humber, Severn or the Scheldt, avoiding shipwreck and without any serious delay or damage to the ship, they would then be qualified to take on whatever the rest of the world had to offer !

THE

MARINERS

NEW KALENDAR.

Containing

The Principles of Arithmetick and Geometry; with the Extraction of the Square and Cube Roots. Alſo Rules for finding the Prime, Epact, Moon's Age, Time of High-Water, with Tables for the ſame.

Together with

Exact Tables of the Sun's Place, Declination, and Right Aſcenſion. Of the Right Aſcenſion and Declination of the Principal Fixed Stars. Of the Latitude and Longitude of Places. A large Table of Difference of Latitude and Departure, for the exact Working a Traverſe.

ALSO,

The Deſcription and Uſe of the Sea-Quadrant, Fore-Staff and Nocturnal. The Problems of Plain-Sailing and Aſtronomy, wrought by the Logarithms, and by *Gunter's* Scale. A Tide Table. The Courſes and Diſtances on the Coaſt of *Great Britain, Ireland, France,* &c. And the Soundings coming into the Channel. With Directions for ſailing into ſome Principal Harbours.

BY

NATHANIEL COLSON, *Student in the Mathematicks.*

London: Printed for *Richard* Mount, at the *Poſtern* on *Tower-Hill.* 1716.

Where you may have all Sorts of Mathematical and Sea-Books.

The fly-leaf from just one of an array of books available from the mid-17th Century dealing with techniques - old and new - of maritime navigation. These books seem to have widely plagiarised but hence relatively inexpensive to buy and as can be seen, contained a wide range of useful information for a navigator to consult. This particular compendium was printed in London in 1716 and seems an update of previous works and includes advertisements for same. *(Copyright National Maritime Museum, London)*

Various instruments, tools and a selection of printed information on how to navigate were available for calculating a position at sea, though as previously stated they all involved the possibility of human error and all depended for accuracy on the degree of personal skill and experience in both the *'dead reckoning'* and *'observation'* methods. The simplest navigational instruments aboard a ship in the late 17th Century were the magnetic compass and the log-line, with any changes in wind or speed course recorded on deck using a 'traverse board'. This simple wooden instrument had small concentric holes drilled in a representation of a compass rose into which eight tiny wooden pegs could be successively inserted at thirty-minute intervals to record the compass course ; another set of small holes and pegs in a different part of the traverse board could be used to record the speed of the ship. Though the traverse board was still in recorded use in the latter part of the 19th Century, with the increase in literacy amongst seamen during the 18th Century this information could instead be 'chalked up' in simple figures on a slate kept at the helm or written down in a notebook.

To calculate the speed of a ship in the original form, a piece of wood - a log - was thrown over the bows and the time it took to pass the stern of the ship noted. Using the known length in paces (or later, feet) of the ship a speed in nautical miles could then be calculated through a simple equation *(see Part Two).* However, the daily wastefulness of valuable firewood would have been noted and this method was developed as a log-line was later attached to the wood and measured off by tying knots placed at seven fathom intervals (42 feet) and sometimes including coloured ribbons placed at every *fathom* (an ambiguous maritime length of distance based on the average distance between a seaman's fingertips measured within his outstretched arms, later standardized at six feet) and a similar but more accurate calculation made using the time it took for a 30-second sandglass to empty, counting the knots as they passed and then hauling the log-line inboard when the sandglass emptied. Daily records of all these readings were made using the traverse board or written in a 'log-book' and as every schoolboy used to know, the speed of one nautical mile per hour through the use of this log-line later became known as a *'knot'.*

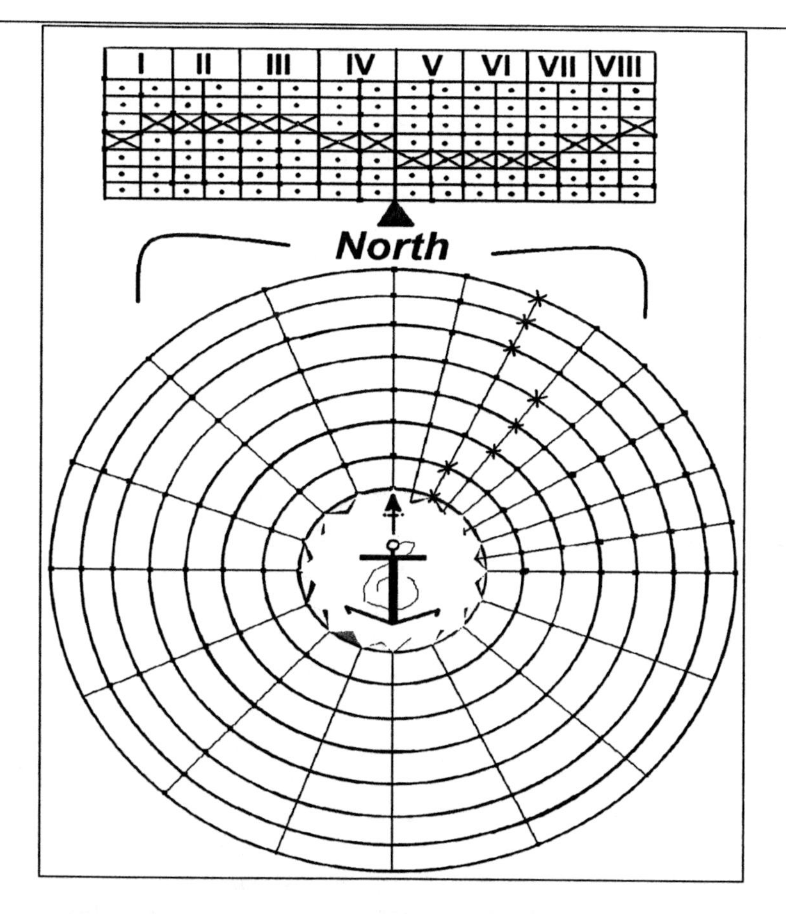

The Traverse Board ; small wooden pegs have been inserted into the holes marked with an X during an eight-hour watch aboard a ship. The top section shows speed, with an average of just over 3 knots in the first four-hour watch increasing to an average of just less than 5 knots during the second four-hour watch. The lower section denotes the compass-card and shows a change in course from *NE by N* to *NE* and a return to *NE by N* halfway between the two periods, probably due to a slight change in wind direction. The principle is shown here but the size and shape of the actual Traverse Board took several different historical forms, left plain or adorned by being carved or painted.

The log-line was then kept on a large free-running wooden spool and the line lowered or cast from the stern of the vessel, weighted at the seawater end with a triangular wooden fitting which became known as the 'chip' which gave drag in the form of a sea-anchor so the line wouldn't be pulled along by the ship and the log-line could pay out freely from a wooden spool. After a new calculation by mathematicians the relationship between Time and Distance was further improved and later it became the norm aboard ship for the 'chip' log-line to be knotted at an interval of 47 feet 3 inches as that distance in a ratio between knots and a nautical mile had now been discovered to be the same ratio as 28 seconds to one hour - the number of knots on the log-line passing through a seaman's hands after the casting of the log-line in 28 seconds was *exactly* equal to the ships' speed in nautical miles per hour. When later hull and sail-plan designs meant a ship could make a far higher rate of knots through the water, to avoid using a far longer log-line a 14-second sandglass was used as the time-keeper and the number of knots in the existing line which passed over the stern during this period was simply doubled. If you aren't in any hurry, the speed of a ship can be controlled by adjusting the sail-plan - many merchant vessels and warships *'made snug'* and reduced speed at night by taking in sail - but as any ocean-going seaman will still tell you, they all prefer a *'fast passage'* as the less time spent at sea the less likelihood of encountering any of the associated hazards in being lost from a variety of reasons such as shipwreck, sickness, enemy action, starving to death or simply falling overboard. As ships aren't equipped with brakes, the log-line or a similar length of 'lead-line' (sometimes known as the 'fathom-line') knotted at intervals of one fathom could also be used by attaching a heavy plug of lead to the end to find the depth of water beneath the keel of the ship to avoid running the ship aground in approaching a coast - day or night - in conjunction with a masthead lookout having a good pair of eyes and ears. A 'leads-man' - or two or more for safety's sake - would secure himself to the bows of a ship or in boat in some way forward of the ship and 'cast' the lead forward into the water, paying out the line until it stopped and reading out the 'knots' which had passed through his hands or as he recovered the line would shout the depth of water to the ships' master at the helm. A shallower depth often - but not always - meant approaching land or a potential reef ; six fathoms was considered pretty safe off a coastline but a sudden shortening to four fathoms would cause concern and a likely change of course or the anchor dropped as even a small trading vessel or a warship of about 250 tons might draw 20 feet of water from the water-line to the keel. Placing wax or tallow in the hollow base of the lead weight on the end of the line would indicate a rocky or a sandy bottom when retrieved hence the best place to drop an anchor, and it is recorded that in poor visibility or fog some very experienced mariners could tell exactly where they were when their ship was *'standing off'* a known coast just by examining what type of sand or broken shell had adhered to the tallow. The use of the modern term *'swinging the lead'* emanates from a seaman employing various means of avoiding hauling the heavy seven-pound lead weight back up from the bottom of the sea any more than they have to - and the maritime terms *'plumbing the depths'* and *'keel'* both have an origin in Latin, as does the term *'lateen'* sail.

The iron ring on the circumference of the maritime astrolabe was used by the observer to hold the instrument for use ; the central apertures were intended to cut down on wind resistance and the base remained heavier in an effort to make holding the astrolabe steadier on the deck of a ship. The astrolabe shown here is made of gilded bronze and measures seven inches in diameter. First recorded for maritime use in the 1400's, Spanish and Portuguese ships in particular appear to still be using the maritime astrolabe in the early part of the 18th century. The mariners' astrolabe can also be used 'in reverse' by using a published nautical almanack and setting the altitude angle at that time of day or night of a star or planet to a compass bearing in order to find it in the heavens. *(Copyright National Maritime Museum, London)*

The maritime astrolabe was developed from an astronomers' instrument firstly as a hand-held simple brass quadrant fitted with a plumb-line used for the angular measurement of stars. Aboard a ship, the wind often disturbed the plumb-line too much for an accurate reading to be taken so the maritime astrolabe was developed to find latitude by the alignment of an adjustable centrally-fixed bar fitted with two apertures in a fore and rear-sight : the astrolabe was held by the observer in one hand and adjusted by his other until the sunbeam from the higher aperture aligned with the lower aperture - when correctly aligned to the sun a reading could then be taken off the scale along the circumference using the pointers and a latitude then computed. The principle of the astrolabe and the magnetic compass appear to be jointly recorded during the days of the Crusades - with allowance for European religious bias made in ambiguous early descriptions - and thought to have their origins in the Arab world - though the Greeks and Romans are said to have possessed similar instruments but like Arab sea-farers, relied mainly on other methods of astronomical observation of the sun, the moon and several stars for navigation, especially Polaris (The North or Pole Star). In the last quarter of the 17th Century it was generally thought by navigators that magnetic compass deviation followed the same north-south course as meridians of longitude, as skilled navigators could use this magnetic variation to estimate longitude - though before 1701 John Flamsteed, then the Astronomer Royal, along with Edmund Halley, the famous astronomer, both had their doubts ; though both men published their findings at this time the problem of magnetic variation was partially solved after 1810 by the use of *Flinder's Bars*, a special fitting of iron bars attached to the ships binnacle which didn't eliminate but helped correct the magnetic variation at sea on a ships compass almost anywhere in the world. Before then, navigators would have to allow for magnetic variation in their daily reckoning and computations and it was usual for the master or navigator and the helmsman of a ship to remove their knives and any other metal objects before *'wiping the slate clean'* and using an azimuth compass to show the difference between true and magnetic north to establish the correct course as a compass bearing.

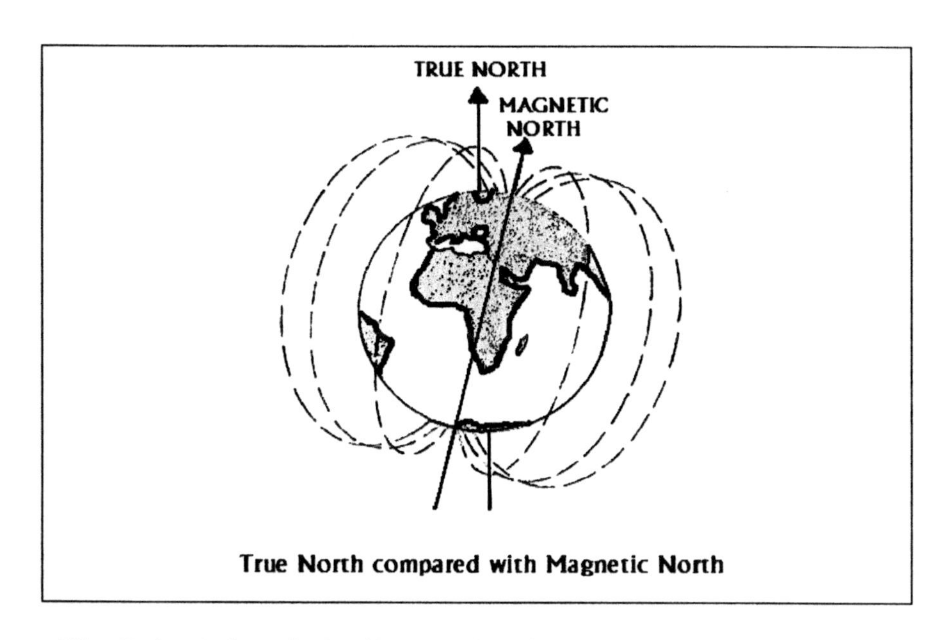

True North compared with Magnetic North

The Principle of Finding Latitude
In the diagram below, point **A** is a navigator aboard his ship. If the sun is south at Noon he is on a latitude in the northern hemisphere ; if the sun is north at Noon he is on a latitude in the southern hemisphere. The distance between **B** and **C** is a constant 90 degree angle from the horizon to a point directly above the observer. The angle formed by **B** is the **zenith distance** angle to a point directly overhead from the observed position of the sun at noon as shown. The angle **C** is the observed **altitude** of the sun above the horizon. The angle **C** in degrees gives the navigator the altitude of the sun (as the sum of the angles of **B** plus **C** will always equal 90 degrees). When the angle **C** is subtracted from the 90 degree right-angle, a latitude of the position **A** can then be computed by the navigator **after** adjusting the angle of **C** for the changing **declination** of the sun throughout the year - the angle measured from the centre of the earth to a point north or south of the celestial Equator between the Tropic of Cancer and the Tropic of Capricorn, an angle of just under 47° through which the sun travels north and south through June to December and back again between the two 'solstices' marked by these two tropics both gaining and losing altitude, all because of the 'tilt' of the earth The seasonal adjustment required for addition or subtraction for **declination** would be gained during the last half of the 17th Century from nautical almanacks which were published regularly and widely

available). The end figure will give the latitude at **A** (but note that making any such observation when close to the Equator is difficult).

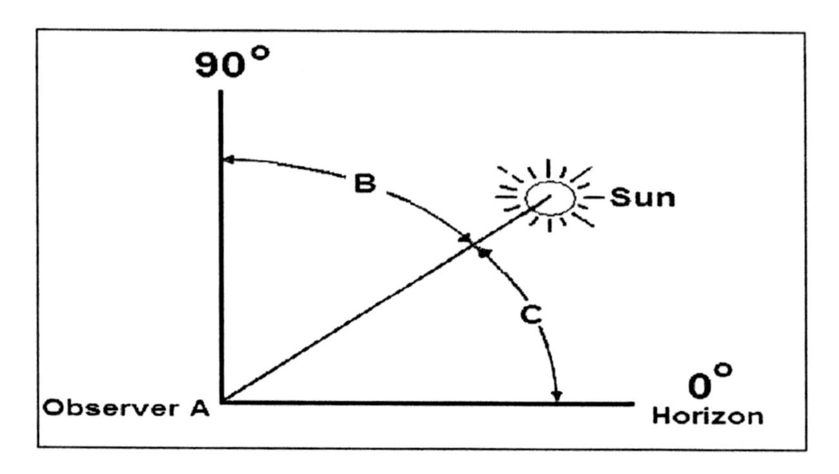

The end figure can then be checked against our navigator's *'dead reckoning'* position aboard his ship - if he has an accurate chart he can plot the position on it and use it to set or change a course. For example : a navigator aboard his ship off the coast of Biafra in West Africa at the point **A** uses a navigational instrument to make an observation of the sun at noon when the sun is directly overhead and due south. When the instrument used in such an observation above but very near the Equator gives the angle **C** as 87 degrees, by subtracting that angle from 90 degrees it leaves 3 degrees (North) as the zenith distance. By taking this figure and correcting it for the changing seasonal declination of the sun in the sky, the computed figure gives the latitude of **A** as just under 3 degrees. As a nautical mile equals one degree of arc along a meridian at that latitude, the navigator can calculate his ship is just under 180 nautical miles to the north of the Equator *(the calculation being 3 degrees of latitude multiplied by 60 minutes = 180 nautical miles).* Four weeks later, the ship our navigator is aboard is now 'somewhere' in the Atlantic at a point east of the Caribbean Sea, having sailed there from West Africa using the 'trade winds' in making for Barbados which he knows lies roughly at a latitude of 14 degrees North of the Equator. After he makes his observation of the sun at noon, he calculates the latitude from the zenith distance, subtracted from 90 degrees and adjusted for declination which results in a figure of just over 13 degrees North. The navigator now knows he has to change the course of the ship and steer more to the north to reach the latitude of 14 degrees, then change the course of the ship again and sail due west to reach Barbados. Making the same observation below in reference to arriving off his home port of Bristol in England, our navigator would obtain the figure of roughly latitude 51 degrees 30 minutes North.

The maritime astrolabe though cast from bronze or brass also remained susceptible to the effects of strong wind on the deck of a ship, making the instrument awkward to remain stationary. The astrolabe continued in maritime use into the 18th Century - especially aboard Portuguese and Spanish vessels due to the instruments ability to be more accurate in observing altitudes in excess of 60 degrees - though the alternative of a far cheaper and more accurate instrument known as the Cross-staff or Forestaff as a means of measuring the angle or height of the sun above the horizon at noon was in maritime use before the year1600. Using the Cross-staff, the staff itself was held in one hand against the eye whilst the other hand adjusted one of the four selected 'crosses' or 'transoms' chosen by the observer along the axis of the staff until two sights were aligned on the base of the sun and the horizon when a reading was then taken from inscribed graduations on the relevant side of the staff *(see Appendix 2).* Though cheaper, handier and more accurate, the inherent problem with the Cross-staff was that to take an accurate reading the observer had to look in two different directions at the same time - one of which was directly into the sun, a fact that is traditionally said to have led to many experienced navigators becoming blind ; and at altitudes higher than 60 degrees, the observer in using a cross-staff rather than an astrolabe found it difficult to 'blink' the eye the distance of the angle between the top and bottom of the cross in order to make the necessary adjustments. To deal with this unfortunate aspect, an instrument known as the Back-staff was designed by an English captain named John Davis in 1595 *(who was killed by Japanese pirates near Sarawak in December 1602)* and developments of this instrument by the mid-17th Century became known as The Davis Quadrant or The 'English Quadrant' - though still generally referred to by seamen as the 'back-staff' through an observer using one of these instruments by standing with his back to the sun. The Cross-staff and The Davis Quadrant both carry scales of zenith distance and altitude, marked in degrees. When aligned with either the angle of the sun above the horizon or the shadow cast by the sun, taking a reading from the sum of the two arcs of the Davis Quadrant is the equivalent of taking an altitude reading from the relevant scale of the cross-staff depending on which of the four crosses has been used - with the Davis Quadrant by using the sum of the readings on the two arcs to show zenith distance and subtracting that from 90 degrees to compute the altitude of the sun, from which latitude can be calculated after adjusting for the seasonal variation in the changing altitude of the sun at noon. But - unlike the Cross-staff, the Davis Quadrant could not be used to take angular measurements of the moon or a star. The 'backstaff' principle had developed by the mid-17th Century into The Davis Quadrant which incorporated the best features of both the Astrolabe and Cross-staff by using two scales fixed upon two arcs. More accurate alternative methods - notably the *Spiegelboog* - saw use aboard Dutch ships ; whose seamen in grasping the principle of back-sight also devised one conversion for the cross-staff in the shape of a combination shadow-vane / horizon-vane and another conversion which sent a beam of sunlight from the top of the cross in use - which became known to the English as the *'Dutch Shoe'* - to a scale marked on the staff so that the cross-staff could be used in 'backsight' observation. The principle and conversion of these instruments was so successful that the Dutch VOC 'banned' the use of The Davis Quadrant aboard their ships in 1731 and it had become common before that time for English navigators to refer to the converted use of 'backsight' applied to a cross-staff as 'The Dutch Fashion' *(see Appendix 2).*

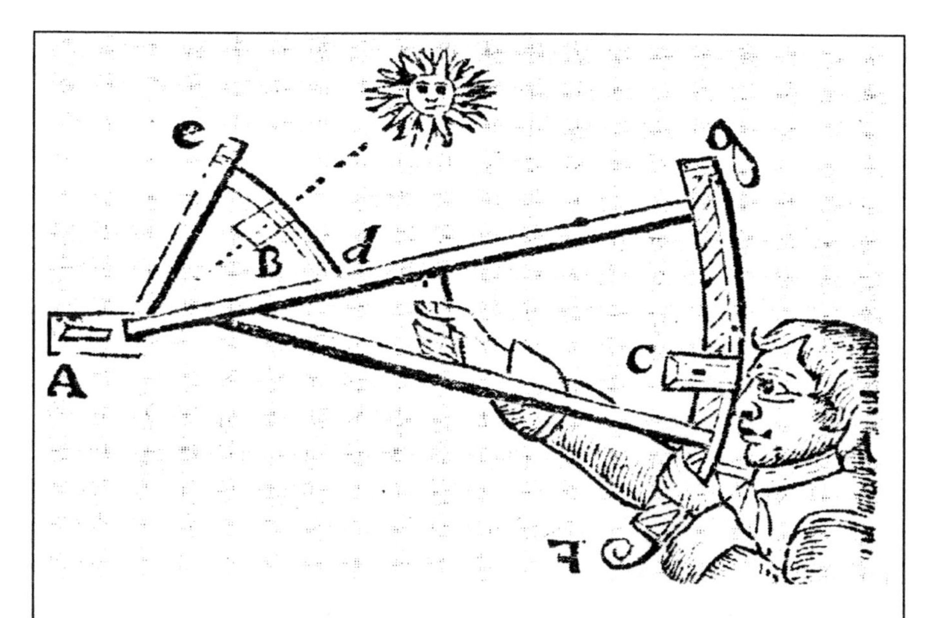

In an illustration originally from *Practical Navigation* by John Seller (London, 1669) a navigator uses a Davis Quadrant to 'shoot' the Sun, a term which was given by seamen in likening the stance of the observer in using both the Fore-staff and this instrument to that of an archer in loosing an arrow. Because the instrument is calibrated to 90 degrees or ¼ of a circle, it was named a 'quadrant' : the instrument uses two fixed arcs, the larger scale calibrated to 25 degrees and the smaller to 65 degrees. When the two vanes B and C are correctly adjusted along the two arcs, the observer can compute the latitude by taking a reading from the scales. When taking such a sight using a 'back-staff' the sun would actually be behind the observer. The letters shown refer to the explanation within the associated book - through an unfortunate mistake by the engraver, the letter F is shown the wrong way around and the Quadrant is shown larger than it actually was. (*Copyright National Maritime Museum)*

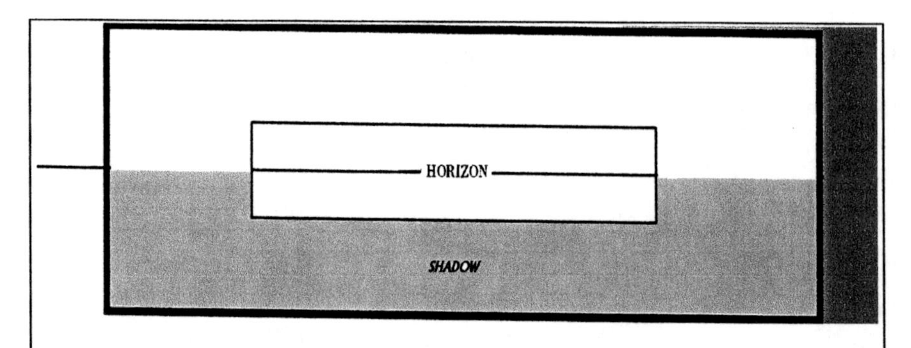

The view of (A) through the sight-vane (C) of the Davis Quadrant as the above observation is made. By adjusting the 'shadow-vane' (B) between (C and D), and the 'sight-vane' between (F and G) the horizon is made to align with the shadow cast by the sun by (B) on the axis shown and a reading is then taken from the two scales on the quadrant (C to D and F to G) ; once the vanes were aligned correctly, the sum of the two readings on the two arcs gave the zenith distance in degrees and allowing for the movement of the ship, a computed latitude should be accurate to within five minutes of arc. A navigator using the earliest form of this particular instrument would require a clear sky to make such an observation but a later development, fitting an alternative 'lens-vane' (B) to the smaller arc of the instrument meant an actual image of the sun could be observed, enabling instruments of this sort to be used on a hazy day or when the sun was 'weak'.

Until the advent of the 'Octant' in 1731 (as the design formed one-eighth of a circle, more properly known as the John Hadley Quadrant) and the acceptance and use of this instrument by the Royal Navy from the British Admiralty in 1750 - but two years before that date, the Dutch - the Cross-staff conversions and the Davis Quadrant remained the main instrument aboard English ships (and for many years to come) in computing latitude. Though not being able to take angular readings from the moon or a star as an Octant could, the Davis Quadrant still remained popular after the introduction of the Octant through being easy to use and despite demanding hardwood construction and careful carpentry, remained far cheaper to buy than the Octant as unlike the Octant no expensive metal or ivory was involved in the construction. The first Sextant - named as the design formed one-sixth of a circle - appeared in 1759 but before mass-production of a much lighter, simplified all-metal sextant fitted with fine-tuning adjustment screws that instrument wasn't used at sea before 1815 but as the price of a sextant fell through mechanical and applied industrial innovation they began to grow in popularity after the first quarter of the 19th century.

The Principle of Finding Longitude

By comparison to Latitude, the rule of Longitude is that four minutes of Time expressed in terms of local and 'standard' time equals one degree of Longitude ; this is based on the fact the earth revolves 360° every 24 hours (360° ÷ 24 = 15°) so for each hour difference in sunrise, noon or sunset between local and standard time that passed on your timekeeper aboard ship when sailing west or east you would have moved 15 degrees of longitude, one degree of longitude being equal to sixty nautical miles at the Equator *(see Part Two)*. As our planet orbits the sun on an elliptical path rather than a circular one, the earth travels faster in summertime than in wintertime and because of this factor an adjustment has to be made (in a similar fashion to declination when computing latitude) by adding or subtracting between a few seconds and 16 minutes periodically to get a truly accurate computation of longitude using Time. The equation to obtain the correction of time was known and the daily adjustment required could be found in a table which was also included in nautical almanacks. The simple '4-minute' rule above sounded quite reasonable to a 'landsman' safely ashore as reliable clocks were around at the time but until the later part of the 18th Century no reliable *maritime* chronometer existed. Accurate mechanical clocks with moving parts designed to reduce friction all depended on a carefully balanced pendulum for accuracy - though accurate enough to use on land, but at sea even the slightest movement of the ship affected the pendulum and rendered such clocks useless. Small pocket watches were extremely expensive and were neither mechanically sound or rust-proof enough timekeepers to withstand the rigours of a long sea voyage and keep accurate time. Before the advent of mail-coaches in the latter part of the 18th Century, even local time in England was not standardised and there was nothing to set your watch or clock beyond observing noon each day and using a local sundial - prior to the stage-coach, the time on a pocket watch carried by a man in Norwich (east England) was always fast at Noon compared to Noon on a watch carried by a man in Bristol (west England), but in using a sundial to set the time both watches were correct - but if the two men met the same day in Oxford and compared watches, they would show different times as one would be slow and the other fast according to local time there. Computing an accurate longitude at sea was proving to be an obstacle to expanding overseas trade through exploration - and the safety of a ship at sea - as even if you did open a new market overseas on a voyage of discovery or simply by accident, unless you were very clever and skilled in astronomy and mathematics and had the required instruments aboard you could not accurately fix a position to report back home to be placed on a chart or map - should such a map actually exist there - so the risk was that you or a fellow navigator would not be able to find the same spot again on a return voyage. Strangely, the British Admiralty of this period - unlike their Dutch, French and Spanish counterparts - created no hydrographic facility or department to create or update charts, leaving that before the mid-18th Century to commercial individuals like the industrious London cartographer Herman Moll, who in 1711 sponsored by The Royal Society and the Admiralty drew the first detailed chart of the Atlantic and Pacific coasts of South America for use aboard British ships - but in this and an updated chart of 1719, 'California' is still depicted as an island. In terms of drawing accurate maps, this meant using as many confirmed reference points as possible. The globe required to be split into equal segments west and east of a standard reference point by both latitudinal 'parallels' and longitudinal 'meridians' - Gerard Mercator, a scientist in Germany, created this facility in the 16th Century and as the sun was on the meridian of Greenwich at noon, the standard 0 degree meridian was set for British cartographers and hydrographers in 1675 with the building of The Royal Observatory at Greenwich, London. The British 0 degree Prime Meridian didn't become an international standard until 1884, with France still using a Prime Meridian which passed through Paris until 1911. Having set your prime meridian, an accurate timepiece could then be used to measure the difference in local and standard time at noon to calculate longitude west or east of that reference point to within a single degree and after perhaps a voyage of over a year when the ship was then on the other side of the world - but the only accurate timepiece available at the time was the observed movement of Planets and the Stars. The full story of how calculating accurate longitude mechanically was achieved is beyond the scope of this introduction as although the quest became 'official' after the passing by British Parliament of *The Longitude Act of 1714* which offered a maximum reward of £20,000 to anyone who could devise a *'practical and useful'* method of calculating longitude to within half a degree, it eventually took an English genius named John Harrison his lifetime to crack the problem (*see Part Three*). Calculating or computing an accurate longitude in the early part of the 18th Century remained firmly within the realms of astronomy in observing the moons of Jupiter but required a method and instruments such as a powerful telescope all designed to be used on *terra firma* and wholly impractical for use aboard ship. As charts did exist in this historical period, it placed calculating longitude at sea firmly into the *'dead reckoning'* aspect, adjusted when sighting *'known land'* which could then be calculated in terms of latitude in order to plot a position to be able to set a new course for another destination.

'Local Time' aboard ship

Rather than being Midnight for a 'landsman' ashore, local time aboard ship began at Noon and fixed by the highest point the sun stands in the sky in terms of angular measurement - when the base of the disc of the sun is at the greatest angle above the horizon seen from the deck of the ship using the instruments described. Local time aboard ship for the purposes of 'standing a watch' was kept by the regular turning of thirty-minute sandglasses and the ringing of the ships' bell. Sandglasses could be used to keep track of 'standard' time from a port of departure but this method was generally found to be impractical and it was commonly held at this time that once 'standard' time was lost aboard a ship at sea for any reason, longitude could not be recovered until the ship returned to the port of departure or arrived at a destination which had a doubly-confirmed longitude. To facilitate the day-to-day handling of the vessel, the ships' crew would usually be split into two 'watches' ; these watches would be in command of one or two of the ships' officers and be divided into the 'forenoon' and the 'afternoon' watch. Using a thirty-minute sandglass, a change of watch would occur at each striking of the ships' bell at the *'eighth turn of the glass'* - that is to say at four, eight and twelve o'clock for each one of the two parts of the ship-day (the latter of which obviously included night-time and members of this watch would be responsible for tending to the ships lanterns during this time). The afternoon watch was also split into two 'dog watches' - between noon and 2pm and between 2pm and 4pm - which resulted in different watches taking turns at the most awkward time between midnight to 4am (known on an English ship as the 'graveyard watch' from the duties of a church sexton ashore). The 'dawn watch' was one of the most important as an unexpected or potentially dangerous situation might be revealed as the sun rose over the horizon and daylight increased, requiring an immediate correction to be taken. *'Watch on Watch'* was a navy system whereas circumstances meant the above principle was applied but the watches would change every four hours ; a good idea in adverse conditions but used as a standard many officers considered 'watch-on-watch' to be particularly tiring to the crew. As the officer of the replacement watch came on station, he would receive the course set by the Master from the previous officer of the watch and his helmsman, be given a brief report from his predecessor about the ships current progress, the ships speed, the prevailing winds and the weather before changing the helmsman for one of his own watch. The new helmsman would then *'wipe the slate clean'* kept at the tiller or wheel and write upon it the present course he would be ordered to steer by the new officer of the watch. All the records from the previous watch concerning course, speed and notes on the weather made every thirty-minutes at the turning of the sandglass during a watch and recorded on the traverse board or chalked up on the slate would be taken below at the end of the watch and 'logged' for use by the Master in 'dead reckoning'. The crew would eat their main meal of the day during the noon watch, prepared in advance by the ships' cook. Important members of the watch - the masthead lookout, the helmsman and his mate and any seaman forward using the 'lead-line' - might be changed every hour by their officer during his watch to relieve physical strain and boredom to ensure efficiency. Though some ships' officers such as the Captain, Master, Boatswain, Gunner, Carpenter and Surgeon did not stand a watch, if the call or whistle for *All Hands on Deck* was heard every member of the crew would turn out to assist, whether it be manning the capstan to raise several tons of wet cable attached to the anchor, an approaching squall requiring an immediate speedy change to the sail-plan aloft or a sight of a 'strange sail' which in this period could mean welcome news from fellow sea-farers and a chance to send any mail back home, or possibly unwelcome news from the same source about another one of the many European wars which regularly broke out between 1650 - 1800 and the ever-present threat from enemy warships and privateers sailing under a *'Letter of Reprisal and Marque'* and of course, pirates ; three more occupational hazards for the poor navigator to try and take into account when setting a course.

In trying to understand the difficulties of these historic sea-voyages in order to recreate them - on paper, anyway - I first had to define the rules and guidelines governing maritime navigation in order to go on to recreate the Early 18th Century maritime character of a *Ships' Artist.*

The author outside a recreated 18th Century chandlers shop demonstrates the principle of The Davis Quadrant in finding latitude. The instrument was commonly called a 'backstaff' as from around the year 1600 it had been noted by seamen that a navigator in using such an instrument aboard ship rather than a Cross-staff now stood with his back to the sun to make an observation. *(Courtesy of Hartlepool Maritime Experience)*

PART TWO

RECREATING *THE SHIPS' ARTIST* THROUGH PRACTICAL INTERPRETATION

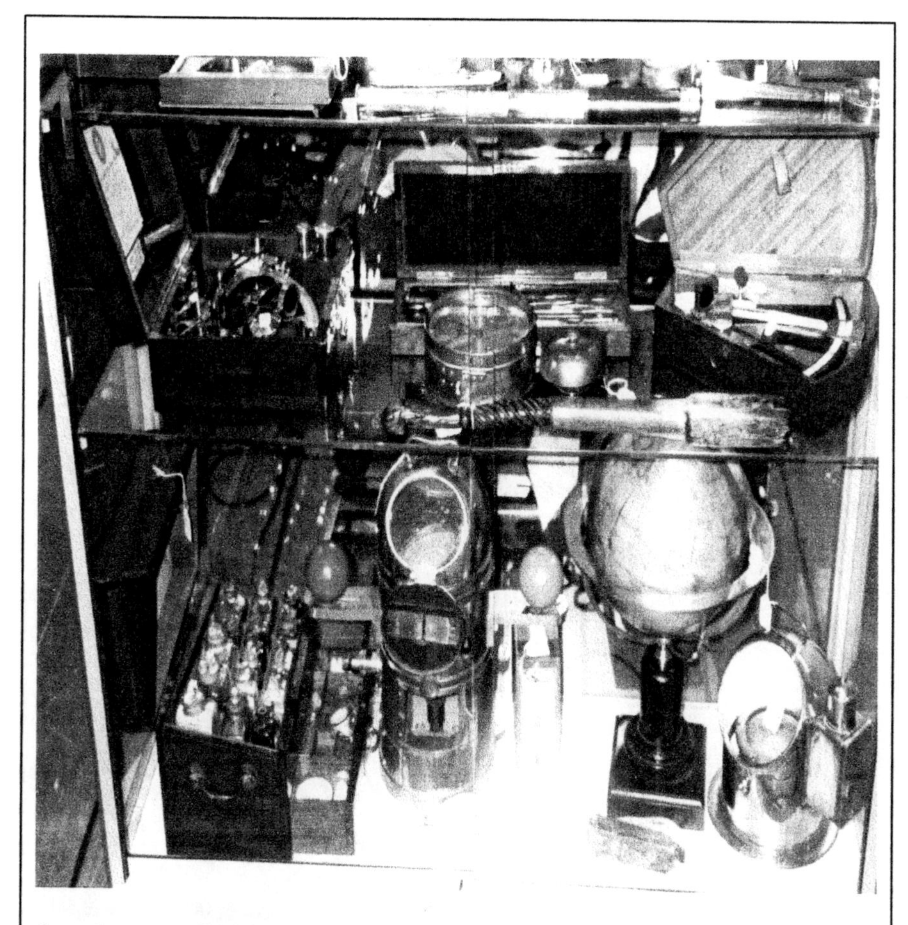

A private collection of memorabilia commemorating *'The Age of Tall Ships'*. Most of the items on display here are 19th and 20th Century in origin but the Octant in the wooden case (top right) and the Surgeons Chest (bottom left) are 18th Century in origin. Access to private collections such as this is very useful when recreating period items as museums often have a copyright policy forbidding any examination or temporary loan of their exhibits for reproduction purposes. *(Courtesy* HM Bark Endeavour *Stockton)*

BASIC NAVIGATIONAL TERMINOLOGY

No matter what his standard of costume, no recreated 18th Century seafarer would be complete without muttering - as an introduction, at least - some of the *'unintelligible gibberish'* always mixed with a selection of *'foul oaths'* overheard and commented on by several 'landsman' of the period as being common amongst mariners. Due to their habit aboard ship of making themselves heard above the prevailing din of the wind, their messmates chatter, gunfire and the usual creaks and groans of a ship in motion, seamen got into the habit of speaking loudly even when ashore and no self-respecting period seaman aboard ship - or ashore for that matter - would use the words downstairs, upstairs, left, right, front or rear. The same applies to navigation and in regard to *Position, Distance, Speed* and *Direction* - the four aspects required to successfully navigate at sea - these terms are shown below :

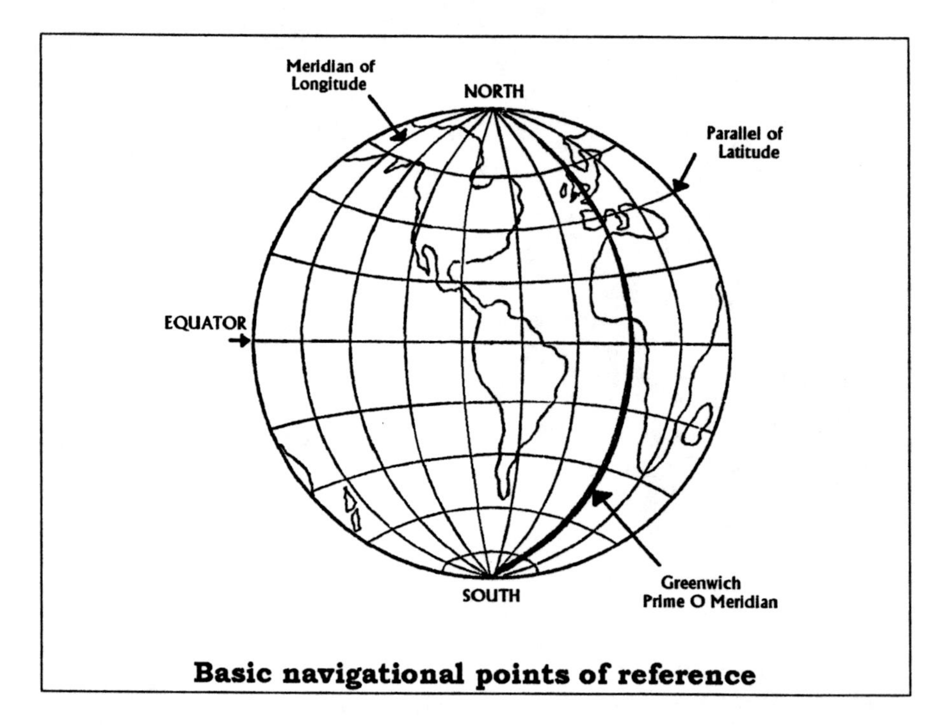

Basic navigational points of reference

Position

Latitude is the angular distance measured along a parallel from 0 degrees to 90 degrees both north and south of the **Equator.**

Longitude is a position on a meridian from 0 degrees to 180 degrees east or west of the 0 degree prime meridian which for British ships in the early 18th Century generally passed through Greenwich, London.

Latitude and Longitude are measured in degrees, sub-divided into minutes, and minutes were later sub-divided into seconds. There are 60 minutes in a degree and 60 seconds in a minute *(a period navigator will seldom use seconds).*

Distance

The unit of distance is the **sea-mile** or **nautical mile** and can be measured on a chart as **one minute of arc measured along the meridian at that latitude** and is 1852 metres or 6076.12 feet. A nautical mile averages roughly 1.15 statute miles but the statute mile of 1760 yards is today never used in nautical navigation - but this *was* in use in the early 18th Century as you can see from period charts ; in addition to miles, distances were often quoted in 'Leagues' which for the English measured three statute miles but the distance varied for Holland, France and Spain. Distances are often quoted in period accounts by using the only other common reference available to seamen ; a *'cannon-shot'* (variable, but over 200 metres), a *'musket-shot'* (about 100 metres), *'half musket-shot'* (about 50 metres) or *'pistol-shot'* (about 25 metres) based on an observation or assumption of how far projectiles from these would effectively carry. A **cable** is one-tenth of a nautical mile *(approximately 600 feet or 200 yards).* A **fathom** is six feet, usually used in terms of depth of water beneath the keel but also from sea-level and as previously stated was a 'standard' derived from the ambiguous maritime measurement of an average length measured from the tips of the fingers between the horizontally outstretched arms of a seaman.

Speed

The unit of speed is the **knot** ; a velocity of **one sea-mile per hour**. The formula to calculate speed for 'dead reckoning' purposes is :

$$\text{Speed } \textit{(in knots)} = \frac{\text{Distance } \textit{(in sea-miles)}}{\text{Time } \textit{(in hours)}}$$

The use of speed in 'dead reckoning' is used as follows, for example ; a ship sails due east at a speed of 6.1/2 knots for 3.1/2 hours. What distance has it travelled on that compass course ? *Distance is 6.5 multiplied by 3.5 = 22.75 sea-miles.*

Direction

The **course** is the direction from one position to another. A **bearing** is the direction of an object from an observer with reference to north. Direction in this historical period is measured in reference to north and is expressed in terms of course and compass bearing and not by a bearing given in degrees. Bearings as we know them today were not in use in the early 18th Century as the compass card did not have a 360 degree notation but a basic 32 point notation on the 'compass rose' placed at divisions of 11.1/4 degrees, later subdivided into 'quarter-points' of 2 degrees 48 minutes and 45 seconds which every seaman would be called upon to memorise in an exercise known as *'boxing the compass'.*

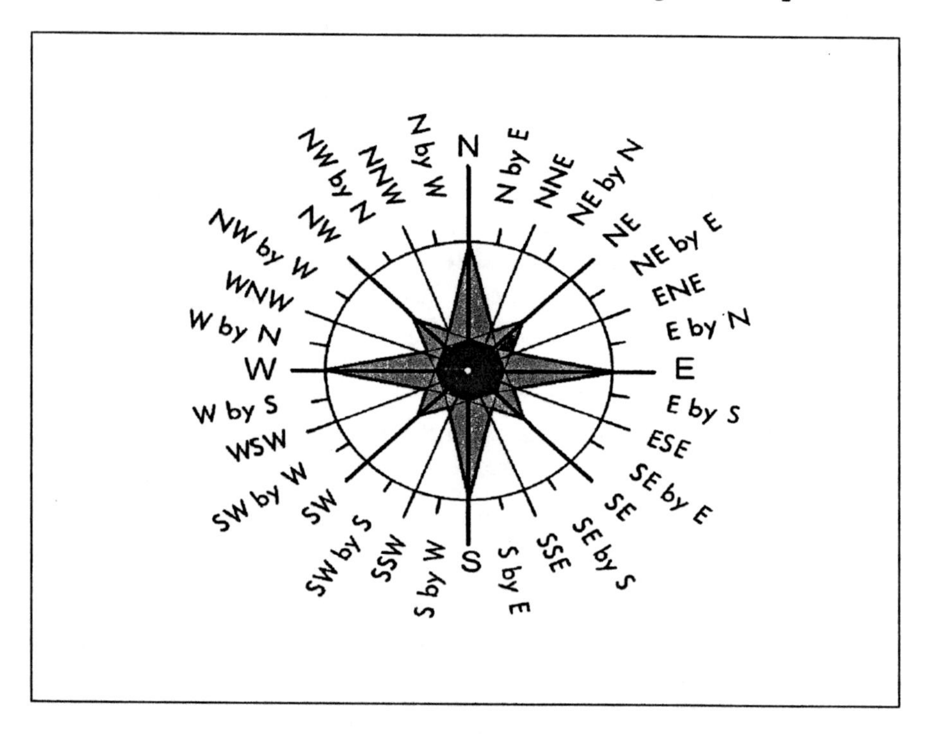

Any required sub-division of the basic 32 points of the compass card for early 18th Century directional purposes could be made by adding a simpler sub-division ; for example, a course between NE and NE by E could be stated through *"Course NE by E ... and a point North."*

Leeway

There also existed a separate terminology for speed, direction and distance when applied to wind, tidal streams and currents. In calculating the effect of *Rate, Drift* and *Set* in 'dead reckoning' on a ships projected course much would depend on the practical skill and experience of the navigator in allowing for **Leeway** caused by the ship moving sideways on a compass course because of tidal streams, wind strength and direction which like magnetic variation on a compass would all affect actual course and hence, position. The speed *(in knots)* of a current is known as the ***Rate***. The total distance *(in sea-miles)* moved owing to the influence of tidal streams over a period of time is known as the ***Drift***. The direction towards which a tidal stream is flowing is the ***Set***. Traditionally, tidal streams flow *towards* a compass direction but winds blow *from* a compass direction. The degree of **Leeway** would be indicated by the log-line when dropped over the taffrail directly astern of the ship in observing the angle between that of the centre-line of the ship shown by the masts and that of the log-line as the line paid out from the spool : the greater the angle, the greater the **leeway** to port or starboard.

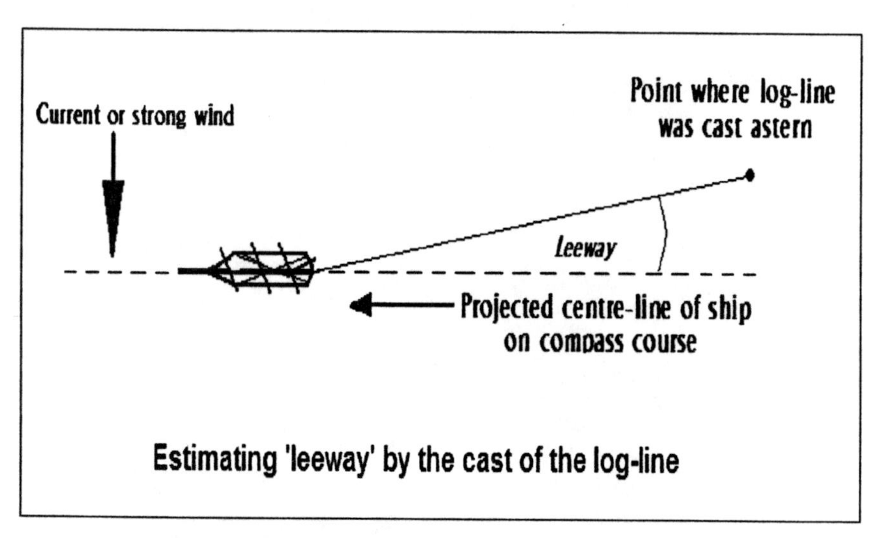

Estimating 'leeway' by the cast of the log-line

Mercator Projections

Accurate charts show an undistorted representation of the shapes of geographical features to give the facility to plot positions in terms of latitude and longitude and to measure direction and distance from one position to another. Distorted charts or maps are useless for navigation ; the 'stretching' causes the true compass direction to alter and mariners using these charts to set a course don't arrive where they want to be. In the 16th Century, as previously described Gerard Mercator had solved this problem by dividing the globe into what became twelve equal *'gores'* or two-hour segments along longitudinal lines of 30 degrees which bisect the Equator at a right-angle and when spread out flat as a map or chart left the compass directions as straight lines. Mercator later bound reproductions of these 'gores' as individual maps into a single folder which he named after a legendary Greek hero who had supported the globe on his shoulders - *Atlas*. Most modern maps and charts for navigational purposes by land, sea or air use the method of undistorted division of Gerard Mercator.

On a map or chart, meridians (longitude) appear as vertical curved lines but parallels (latitude) appear as horizontal straight lines. The meridians at the Equator are drawn as equally spaced straight lines. The simple fact that the world is round means that the most direct course from a position to another position or the observed bearing of an object or point would appear as a curve. To compensate for the straightening of the meridians of longitude and maintain the correct shape of the land, the north-south distance between successive parallels of latitude is increased in the same proportion. The most convenient form of line to use in navigation is a *rhumb line* which cuts all the longitudinal meridians at the same angle on a chart as all the meridians of longitude cross the parallels of latitude at a right-angle appears on a Mercator projection as a straight line. The rhumb line differs by so little from a true shortest-distance straight line over distances up to 100 miles that no significant loss of accuracy in assuming all courses and bearings are rhumb lines, which radiate outwards from the 'Compass Rose' shown on most charts. Rhumb lines appear as straight lines cutting all meridians at the same angle. This enables a navigator to *'lay off'* and measure easily a rhumb line course and bearing. *'Great-circle'* tracks appear as straight lines on a sphere or a globe and this facilitates the planning of long ocean-going passages in which the great-circle distance is significantly shorter than the rhumb line. In very high latitudes such as 70 degrees where meridians converge towards the poles - a place where most of our period navigators would never venture - it is not practical to use a Mercator projection owing to the extreme distortion of the latitude scale. The 'Compass Rose' shown on all charts may indicate True North or Magnetic North and radiates *rhumb lines* as directional indicators to a navigator according to the compass card. Using the Equator with each 30-degree segment laid side-by-side, Mercator's original 'gores' show both undistorted land masses and seas and true compass directions.

'Here be Sea-Monsters'

Many 'living history' practitioners and recreated period navigators will already be familiar with the above phrase, traditionally shown by a sketch of a mythical sea-monster on ancient charts where that area was utterly uncharted as a discreet warning to prospective navigators intending to go there ; but as has been explained, disasters occurred even in well-mapped and frequently-sailed seas. Accurate charts were priceless to merchants sending vessels further on trading voyages, and upon their safe return to a home port updated charts fetched a high price from Governments who used them to amend the charts issued to their navies, especially as trading areas steadily expanded east and west into fresh parts of the globe and competition from all other maritime nations was fierce. As will have previously been noted, one buccaneer of the late 17th Century escaped arrest and trial and instead received a royal pardon and a sum of money as a reward in having captured Spanish charts which showed much detail of hitherto uncharted regions of *The Great South Sea* after he astutely copied and 'donated' these charts to the British Crown.

Due to the biblical reference to Jonah, whales were feared by early seamen as potential hazards with other strange aquatic mammals such as the walrus and the hippopotamus. The manatee - glimpsed in the water by short-sighted, drunken or sexually-frustrated mariners - gave rise to a similar myth about 'mermaids'. Here, a fearsome sea monster destroys a ship in the vicinity of Cape Bojador off West Africa in one corner of a restored 16th Century Portuguese chart. Such creatures were the excuse superstitiously given by seamen for the non-return of a ship after sailing into uncharted waters. By the beginning of the 18th Century, belief in sea monsters still persisted as more than plain superstition but in a later chart, an alternative message is given as apparently the same sea-monster is shown stranded on a beach being slowly dismembered by seamen ready to be cooked and eaten !

In order to demonstrate or explain early 18th Century maritime navigation in a 'living history' role, you require a working knowledge of all the above principles - and their future development - and have undertaken some practice along the lines of the simple exercises shown in this book. Once you've grasped the basics, a few props will help you look and feel the part and also help you explain the system and demonstrate how it actually worked. Any display of maritime navigation for 'living history' purposes using the items described below will require a flat table surface but in a tent may unfortunately be subject - like the weather deck of a ship - to the nuisance caused by wind and rain. In such circumstances by using a pair of flintlock pistols as paperweights on my chart they were found to distract my younger visitors from what I was showing them and in future I used small, flat pieces of lead sheet instead. Performing a navigational display in a tent or aboard ship, good lighting is essential but be aware of the potential hazard in using candles or any other flames. You *might* in your introduction keep the degree and depth of mathematics down to a minimum leaving how to calculate in detail to a short explanation on a sheet of paper or using a slate and chalk later if anyone wishes to know (and some visitors will ask you to demonstrate this). For my own straightforward 'have-a-go-yourself' navigation in a challenge to *'Find the Buried Treasure'*, I use two charts with a sheet of basic data ; one chart is large-scale of my own design of the Caribbean Sea and another a copy of an original 18th Century smaller-scale Spanish chart showing The Bay of Honduras which rather conveniently charts a spot named *Laguna de Piratas* - 'The Lagoon of the Pirates' - which is a rather good attention-catcher. Once you've plotted your position on the large-scale chart using the data, using the clues offered this becomes the spot to set a course towards using the latitude, longitude and other maritime data contained on the chart. You can offer as many or as few clues as you wish to suit the intention or the age of the audience ; in past exercises - which were quite entertaining to all concerned - we have had ships sailing in the wrong direction, ships running aground on islands and sandbars, ships running out of water (the captain then having to handle a mutiny), being fired on by shore batteries, captains losing track of the associated time-scale and attempting disastrous manoeuvres at night -

and some captains successfully translating all the given data and reaching the 'treasure' - a bag of chocolate coins !

The Logbook

Logbooks were available to purchase through a ships chandler or issued by the East India Company and the Royal Navy to their vessels. Basically, it is a stoutly-bound book with a good thickness of many pages in which the navigator would faithfully record all his observations, readings and calculations on the left-hand page and enter relevant comments on the right-hand page. Sometimes - with or without requiring the captain's permission - mariners such as the navigator kept an associated personal journal. Later in the 18th century, printed 'forms' and entire logbooks came ready lined and divided into columns but you can always rule your own columns or sub-divisions along the lines of the example shown using a 'rolling-rule' and a lead pencil. Most surviving logbooks show careful and clear long-hand writing in ink but this was not always the case and some logbooks are casually written in *plumbago* using a 'mechanical' pencil. To save time in writing out many pages of an imaginary voyage for your demonstration, it's a good idea to start out with logbook 'Volume Two' and begin your records on page one with a position in the middle of a sea or in a faraway port which you can show on an associated chart. Logs not unusually carried details of any misdemeanour by a crewman, sightings of other ships and reports of any unusual event of any sort and were often accompanied by copies of any new charts drawn or obtained ; Columbus' log from his first voyage makes interesting reading in this respect especially in the report of several men aboard his ship seeing 'strange undersea lights' at night to the south of Bermuda ? It was not unusual for an Able Seaman to be able to read and write in the period dealt with in this book, but the keeping by crew of personal logs aboard ship was frowned on by the Admiralty - it was contrary to *The Articles of War* in the Royal Navy and liable upon discovery to see a crewman charged with sedition or mutiny. Being able to cut a quill with your penknife and having a 'portable' inkwell are handy assets for any practical period navigator, but you can fudge this and avoid both harmful spillages and not being able to erase a mistake by writing your logbook at home using a metal nib or a fountain pen and for demonstration purposes just make notes on a separate sheet of paper using a plumbago pencil with the explanation to visitors that you will transfer all this data into your logbook 'when you come off watch'. The logbooks and journals of several explorers, buccaneers and privateers later became published in book form ; William Dampier is a noted example and the accounts of Woodes Rogers, Edward Cooke and William Funnell are also good examples as these describe the same voyage from different viewpoints (see *Further Reading*). All Royal Navy logbooks and some of the merchant marine were later required to be handed over to the Admiralty at the end of each voyage to avoid them being published in advance of any 'official' publication.

The Chart

A simple chart is by far the best means of demonstrating and explaining navigation and even a plain sheet of paper can be made to serve using a few tools such as a compass, a parallel rule or a rolling rule, a pair of callipers and a pencil. The 'practice charts' illustrated in this book can be easily reproduced, enhanced and expanded into the second or third versions shown. Copies of period charts are available but are not always the best to use in a 'hands-on' demonstration as some don't show any parallels or meridians at all and depending on which country the chart originated may show a prime longitudinal meridian passing through London, Paris or Cadiz or where a particular voyage began. Period fonts can be downloaded from the internet and replacement photocopies in A1 or A2 size cost a pittance if you stick to monochrome, but remember when designing your facsimile chart by computer, everything on the chart - including the font - will be much bigger upon enlargement at the printers. As with period charts, it is a good idea to leave a good margin around your design to allow for any enlargement difficulties and navigational calculations and scribbles. Having your facsimile charts photocopied onto off-white cartridge paper or card is more expensive but the finished article both looks and feels more authentic. Large coloured charts look impressive but require more work to create and space to use so I use a selection of different sizes to suit the circumstances. If you are doing chart-work on a table, a smooth surface is desirable and a square sheet of 2mm thick leather is handy to cover a ships' table or a similar flat surface having or preventing dints or scratches in order to both prevent any damage and the sharp points of callipers or a pencil puncturing the chart. Facsimile charts are available for mail-order on the internet, but some of these can be expensive for single use. A tube of some sorts in which to store charts to avoid folding can be fashioned from copper sheet using a soldering-iron but a cheaper option would be a disguised cardboard tube or even a 'camouflaged' length of plastic pipe. Large charts for public have-a-go do facilitate the use of tiny ship models (available from model or war-gaming retailers) and you can have several of these on the chart at the same time, introducing the additional threat of 'potential enemy action' for a navigator to consider should two ships meet and one of these turns out to be a pirate !

Journals and Periodicals

Depending on the depth of your pocket, both original and a variety of reprints are available of maritime almanacs, books, lunar tables, journals and relative works. In considering using an original, remember that your available display paperwork shouldn't look 200 years old but be in good condition, showing only recent wear and tear. You can always make your own journals and you might consider a reproduction printed form or two showing when and where you passed your *Masters Examination* and getting someone skilled in calligraphy to fill in the relevant details in *'copperplate'* to impress your visitors! Having back-up paperwork as to your past experience isn't necessary but it is another handy facet to use in a display. You can always re-cover this particular book by using leather or stiff brown cartridge paper in a facsimile fashion to have it close to hand for reference. Period-looking fonts such as *18th Century Book* can be downloaded from the internet onto your computer and be used to create facsimile paperwork.

Most ships had a small, clean and dry area inboard to use as a chartroom for navigational purposes but if not, the large stern cabin would be used. Generally the personal space of the Captain, the stern cabin usually held at least one chair and dining table ideal for laying out charts, and large windows cut in the stern gallery itself admitted plenty of light. The 'anchor watch' of a master's mate and two crewmen seen standing on the weather deck just above the small windows here are where a 'skylight' window to the stern cabin was also commonly fitted. The windows all have shutters to avoid storm damage at sea to the fragile glass panes.

Instruments

Period-looking sandglasses of the correct shape, different capacities and several sizes are available from handicraft suppliers and these can always be customised using bespoke pieces of the frame turned out on a simple wood-lathe for a 'period' look. If you can't find or buy a thirty-second sandglass for use with the log-line you can always shave a small hole in the top of a common 3 or 4-minute bulb with a needle file and remove some sand to suit ; you can also replace coloured sand in a bulb by the same method but remember to use clean, fine and dry sand to refill it before you block the hole. Reproduction directional compasses - both boxed and unboxed, and with or without floating on a *gimbal* - come in many shapes and sizes so try and select one that best fits your selected historical time period and without the modern degree scale on the circumference of the compass card. You can always download a 'compass rose' from the internet and make your own compass and card : a 'slate' and a 'log-line' are two more things you can easily make yourself, but the acquisition of a collection of period instruments largely depends on your budget.

Some useful navigational gadgetry and instruments are available by trolling ebay, antique and collectables shops (even some garage or car-boot sales) but a period-looking range of instruments for beginners can be obtained by mail-order - but remember that most modern reproductions of period navigational instruments are not intended for any serious use. As they are there to be handled by members of the public, I personally have a mix of real and reproduction items, but remember also in terms of making these available to visitors for handling that most of these are easily broken if accidentally dropped or mishandled and a pair of dividers or callipers usually have sharp points. A lump of period-looking 'india-rubber' purchased from a stationer can be used to erase pencil markings or errors, but watch out for the ink-bottle as a spillage could be disastrous in period surroundings. If you are using parallel rules, you can check the alignment of an old chart by 'walking' these over to the compass rose. The historical parameter you set yourself in terms of a cut-off date will denote which instruments you might possess at that time, unless you are presenting a general display of 18th Century Maritime Navigation.

Anachronism Note ... *Modern sextants are fitted with coloured glass eye-shades but if you use the early 18th Century navigational instruments described in Part One of this book and wish to experiment or demonstrate the Principle of Latitude, don't look up into the sun without wearing some form of approved eye protection as though smoked glass is 'in period' it won't prevent damage to your eyes.*

What instruments you use in any 'living history' display depends on your chosen historical 'cut-off' date. This set-piece navigational display by the author set in a chart-room aboard ship shows a general selection of original and reproduction 18th Century navigational instruments. Clockwise from the centre-left : a boxed directional compass, a slate and chalk, a small azimuth compass, a 28-second 'log-line' sandglass, a larger 30-minute watch sandglass, a standing magnifying lens, a rolling rule for use with the logbook, an ivory-handled magnifying glass, small parallel rule, a plain plumbago pencil, a 'mechanical' pencil, a large parallel rule and three different forms of dividers / callipers. The instruments lie on a copy of an original 18th Century chart in the authors collection showing part of the Bay of Honduras. Brass and bronze are favoured metals for use aboard ship - though not as durable as iron or steel, these metals don't rust when exposed to saltwater spray.

A selection of original 18th Century writing implements used by the author in displays. From bottom left ; silver spectacles in a case, a silver pen-wipe and inkstand, a stamp used with printers ink, a pounce-pot, a seal for wax, a *porte-crayon* or 'mechanical' pencil, two cut quills, two glass inkwells and an ivory wafer box, a brass pen-wipe, a Venetian Glass pen, three pens fitted with steel or silver nibs, sealing wax, three penknives, an ivory wafer-seal and a *gratoire* used to scrape away mistakes from paper or parchment. All these items are practical and used for their intended purposes.

'Sea-Chest'

It's a good idea to have a period-looking box to keep all your navigational instruments safe inside, along with a small bag of the requisite cleaning materials for same. Many of these instruments come in their own box, so view what you have or intend to acquire and estimate a size of chest to accommodate them safely. If you wish to accommodate charts, a good size for a chest is just over two and a half feet long, a foot deep and a foot wide, fitted with period handles - but without including storage for charts you can probably manage with half that size and that will fit down and along a ships gangway a lot easier. With basic woodworking skills and tools you can easily make a stout box or sea-chest from 20mm thick fresh or reclaimed pine and using commercial wood-stains 'dress it up or beat it down' to look a little used, perhaps adding a 'broad arrow' to show Government ownership or branding your initials into it with a red-hot poker (or an electric soldering-iron). Don't bother with any dove-tailing - use plain joints, recess the woodscrews then tap in a short piece of dowel into the recess to hide the head of the screw. If you'll be opening the chest in front of your visitors, you might try the trick of lining the interior of the chest with a sheet of reproduction period newspaper, available from several suppliers and also on the internet. To avoid damage to the hinges - should the lid slip and drop back - fit a small length of brass chain as a retainer inside the chest. A range of cheap but sturdy reproduction ironware is widely available to suit the outside veneer of such a chest ; flat lids are best for sea-chests as they will stack safer and if stoutly-made when covered with a strip of padded tarpaulin can be used as a seat. There is nothing at all amiss with having a stout sea-chest with a stepped or rounded lid in this historical period, but these sea-chests are more difficult to construct and depending on your woodworking skill you may wish to seek out a local carpenter to make you one. My sea-chest has 'tapes' crossing the lid, behind which you can tuck papers and documents ; it also has an interior lid which serves as both a carrying-tray and storage for several small everyday objects and a small drawer compartment for cash and valuables. The chest has a flat base to prevent vermin hiding beneath it but you can always make a wooden stand to raise the height of the chest from the deck to avoid bending down too far. For peace of mind it's always worth considering security ; in your absence on deck or in your bunk, fitting an flush interior brass box-lock is best, but a plain hasp and staple with a period-looking padlock will suffice.

PART THREE

Developments in Navigation from 1492 - 1800

Though development was regularly interrupted by wartime, nautical charts and global maps steadily continued to improve after 1700 through greatly increasing trade and *'voyages of discovery'* - especially those in mid-century commissioned by England and France - and combined with developments in printing by 1780 increasingly accurate charts were both cheaper and more accessible to all navigators. Improvements in accuracy continued through the introduction of the Octant and the maritime chronometer and with both latitude and longitude able to be computed at sea by a practical method, a graduated scale placed along the edges of the nautical chart enabled far simpler position finding as a navigator could now compare and relate his observations with an accurate chart with increasing confidence that most charts related to what he would actually find when his ship arrived there ...

A Navigational Problem for Christopher Columbus in 1492 ...

At 2.00pm on the 12th day of October 1492, an ordinary seaman named Rodrigo de Triana - the lookout perched in the masthead aboard Columbus' ship - shouted to the deck below him that he had sighted land. Columbus heaved a sigh of relief as he had sailed west from Spain in the knowledge that based on his calculations he should have reached Japan in September. On 7th October he had changed the course of his ship to west-south-west to follow the same course of a flight of seabirds which from an old Portuguese mariner's proverb he believed indicated land in that direction ; but with no land sighted and a further change of course to south-west on 9th October his suspicious crew stated that they would sail for only three more days on that course and demanded that the ship must go about and return to Spain if no land was sighted by the end of the third day. Halfway through the third day - they sighted land. Columbus before setting sail on this voyage had based his calculations - and his future expectations - on longitude, and having reached the required longitude he expected to reach *Cathay* (China) or *Cipangu* (Japan) : but they weren't there. Columbus had actually sighted what later became San Salvador, later reaching Hispaniola (now Haiti and San Domingo) on Christmas Day. As his ship ran aground there, Columbus still believed that he had discovered a quicker route to the Orient and the 'East Indies' lay to the south. Only after three more voyages did Columbus realise that he hadn't reached the Orient at all and had in fact discovered a whole 'New World' ... the Americas. For his calculations, Columbus had used a 9th Century text written by an Arab named Al-Farghani which had been translated into Latin in the 12th Century. In this text, a degree of longitude - using the base unit of the Arab mile of 2165 metres - was calculated to be 66 statute miles or 56 nautical miles (but in reality is nearer to 60). Columbus in error believed the text equated to 'Italian' miles - 1480 metres in length - so his calculations were based on the distance between degrees of longitude being 45 nautical miles so in simple terms, Columbus had under-estimated the size of the planet by 25%. In his second error, Columbus used a work published in 1483 containing a previous calculation stating that Europe and Asia covered 225 degrees of longitude ; Columbus 'updated' this calculation by adding another 58 degrees to include the land area recently discovered by Marco Polo, giving a total of 283 degrees from the west coast of Portugal to the east coast of Japan. In reality this figure is 150 degrees of longitude so Columbus placed China in the position actually occupied by Central America. Basing his course on beginning at the longitude of the Canary Islands, Columbus calculated that sailing 2400 nautical miles due west his ship would reach Japan. The actual sailing distance is 10,600 nautical miles. A cartographer and scholar at the Spanish court named Peter Martyr d'Angheira had correctly calculated the size of the globe and publicly stated upon the return of Columbus to Spain in 1493 that whatever land area Columbus had found in the west it wasn't Japan or China, but in the excitement of the new discovery this statement wasn't seriously considered and Columbus in his hour of triumph was believed. The name *'Indians'* that Columbus used in describing the indigenous population he met after landing has firmly stuck to this day. By the year 1500 - when Columbus was in prison and the fact was dawning on him that although he hadn't in fact discovered a 'quick route to the Orient' but something far more important - he stated that his errors indicated the hand of God in the Bible passage from Isaiah that *'God made me the messenger and showed me where to go'* and subsequently changed his claim to have discovered Japan to a claim he had in fact discovered 'The Garden of Eden'. Whatever Columbus thought or said, Spain was very happy with his discovery and claimed the 'New World' as their sole property - especially after a 15,000 foot-high mountain of silver ore was discovered in Venezuela and again after 1519 when Cortes returned from the expedition to check out reports of a place where pure gold was as common as pebbles in an empire ruled by a king known as *El Dorado* .... literally, 'The Man of Gold'.

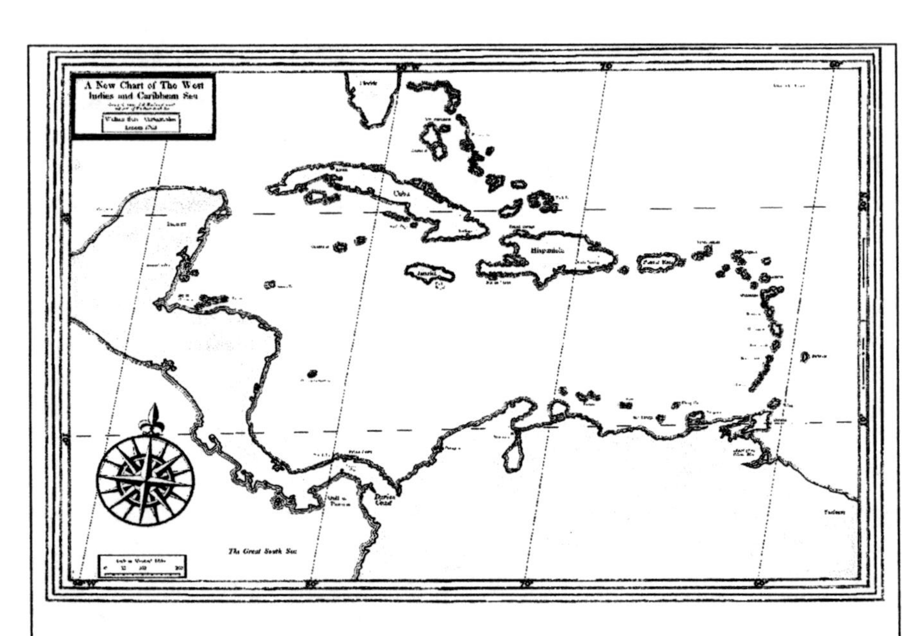

Another development from a basic chart, again designed for easy reproduction drawn by the author using a pantograph for 'have-a-go' and when enlarged to full size makes a useable table-top chart for practice or demonstration (note that for educational purposes this particular chart has a built-in deliberate mistake). Printing on cartridge paper gives a chart a 'period' feel when handled. Copies of period charts are available on the internet from several sources if you aren't able to draw your own (but if you intend to photocopy any period charts, remember they *may* be someone's copyright).

The Navigational Problem at the end of the 17th Century
Finding a destination by 1689 from a ship was resolved though sailing north or south and then by using your instruments, sailing along the latitude of your destination until you reached it. This was still the most popular method, though taking far longer was always judged the safest course. As previously seen, in 1492 Christopher Columbus correctly calculated the latitude and had he correctly calculated longitude he would have eventually reached China - but several months after his ships had run out of food and water had the American continent not been in the way. By 1689, two methods were used by astrologers mapping the earth masses : specifying stars by 'declination' (the star's angular elevation above a plane drawn through the equator - and by 'right ascension' (the angle measured by the time elapsed between the star's passing through the meridian and that of the sun or another star). With the increase in accurate pendulum clocks, a method of astronomy was recommended from 1630 using one of the new design of telescopes and the moons of Jupiter but this was still difficult enough on land. At sea, a ships position in latitude was ascertained away from sight of land in northern latitudes by measuring the angle of the Pole Star above the horizon, as one degree of latitude was judged by navigators equal to roughly fifty nautical miles of distance north or south. In the southern latitudes, the sun was used making some seasonal adjustments for which by the middle of the 16th Century printed read-off tables had been published and were available. Mariners sailing inland seas or coastal waters from ancient times had been able to judge latitude using common sense and experience in reference to the heavenly bodies, the set of currents and the ebb and flow of tides or more recently a variety of evolving instruments such as the astrolabe or fore-staff. Though latitudinal and longitudinal lines began to mark the planet from the third century BC, for a 16th Century ocean-going mariner out of the sight of land to fix his latitude accurately on a chart, he was required to perform a calculation based on a heavenly observation. In the northern hemisphere, the angle of the altitude above the horizon of the stationary Pole Star was commonly used as navigators knew that one degree of altitude was equal to about fifty nautical miles of distance. In the southern hemisphere, the sun's altitude at noon was used with a proportional adjustment for which by

the 16th Century were available in yearly published tables. At the time, calculating or estimating Longitude decreed 'dead reckoning' based on course, speed and taking into account *leeway* as best you could - the accuracy of any estimation varied with the amount of experience possessed by the navigator - but as most Ordinary Seamen remained illiterate before and after the 18th Century was a task left to the captain, master and any master's mate but still involving some estimation and a risk of human error ...

On 22nd October 1707, during a good deal of confusion caused by twelve days of dense fog and mist, five Royal Navy warships became separated from a fleet of twenty-one vessels which set sail from Tangier on 30th September returning towards England and a successful naval engagement and ran onto rocks off the Scilly Islands with four ships wrecked and two thousand seamen drowned. The confusion arose over longitudinal positioning : because of the confusion, Admiral Sir Clowdisley Shovell - the anxious senior commander - had previously called the fleet navigators together on board the flagship who all confirmed the squadron was on a northerly heading safely in open sea off the west coast of France ; the fleet was in fact 150 miles further to the west heading straight towards coastal rocks. An experienced able seaman on board the flagship at the time who had kept his own 'dead reckoning' positional reckoning thought the situation so serious that he approached Shovell and declared the flagship was about to run onto these rocks, an act of 'subversive navigation' being strictly forbidden in The Articles of War which saw the seaman hanged at once for mutiny. Shortly after this execution, the seaman who had just been hanged was proved correct as the rocks were sighted by a masthead look-out - but too late for the squadron to change course and four of the ships ran onto the rocks and sank. Only two seamen survived the disaster and safely reached shore *(traditionally one of them was Shovell who whilst lying exhausted on the beach after the swim ashore was then murdered by a local woman in order to steal an emerald ring he was wearing).* The logbooks of the surviving ship showed the squadron's initial calculation of longitude was wholly inaccurate and further confused by twelve more days of compound error - but - it was said that had it been possible for the initial longitudinal calculation to have been made correctly, despite the fog the subsequent 'dead reckoning' based on an estimation of course, speed and 'leeway' would have placed the squadron in the correct position and approaching the Scilly Isles. Prior to this disaster in May 1678, eighteen of the thirty-five vessels in a Franco-Privateer fleet aiming to destroy the Dutch colony on the island of Curacao by surprise had been wrecked off the coast of Venezuela with the loss of hundreds of lives when at midnight in sailing west with the trade wind they ran in succession onto the notorious six-mile long reef north of the tiny island of Aves de Barlovento, though this reef had been charted and many of the privateers included in this fleet claimed to have good knowledge of those waters so theoretically knew exactly where they were - but in leading the fleet it was one of the privateer ships that struck the reef first. The fleet commander - Contre-Admiral Comte d'Estrees - had thought at the time his ships were safely in open water many miles to the north ; when the Comte was roused from sleep and made aware of the potential destruction to his fleet, he ordered carriage-guns aboard the flagship to be fired as a warning signal to the ships astern but in the darkness this signal was mistaken - four of the French warships thought the flagship had been surprised and engaged by an enemy. In seeing the large lantern still burning at the masthead of the flagship as a beacon, these four ships increased speed and quickly closed on the flagship in support - and drove one after the other onto the reef. Only the rearmost half of the fleet managed to realise what was happening and change course to avoid this disaster but were widely dispersed in doing so. Accounts of the disaster state between 300 and 500 men were drowned in this calamity with 1500 surviving seamen and soldiers becoming stranded on the rocks of the reef above water, with the surviving privateer ships - ignoring the distress of the castaways as they tried to clamber over the rocks and coral-heads to reach the relative safety of a tiny island - only being concerned with standing off the reef in order to collect the large amounts of salvage as the wrecked ships broke up, , then sailing away with their plunder. Three weeks later, a rescue ship arrived to take off the remaining survivors from the island in the reef but the loss of 1500 lives and five of the most powerful French ships afloat in this disaster completely changed the balance of power in The West Indies. Though the two instances quoted here were serious enough because of the large loss of life and vessels, both disasters then necessitated a major re-think of their nations' entire current naval strategy and once again focused the need for more reliable and practical methods of navigation ... especially Longitude.

The Longitudinal 0 degree Prime Meridian at Greenwich
Latitudinal parallels are fixed by nature, being equal divisions of a right-angle of 90 degrees of each hemisphere north and south of the equator. The origin of the 0 degrees longitudinal meridian fixed at Greenwich in London goes back to the founding of The Royal Observatory at Greenwich in 1675. Still having a choice of reference based on English, French and Dutch cartography, mariners up to 1767 commonly gave their longitudinal position before then as so many degrees east or west of the nearest meridian on the chart they happened to be using but most often using the longitude of their port of departure or that of their intended destination. The popularity of the *Nautical Almanac* - printed tables in book form published in London by The Royal Observatory - from 1765 until 1811 gave astronomical information based on that spot as the prime reference ; hence the beginning of general acknowledgement of the 0 degree meridian in Greenwich, seven statute miles from London. This continued through the introduction and development of chronometers as navigators had to set their timepieces before they could be used at sea and continue to make lunar observations on an average of once every month to confirm these chronometers were accurate and hence began to compute their position so many degrees east or west of the Greenwich meridian. Having the popular *Nautical Almanac,* navigators simply used the relevant page in the *Almanac* to check their English-made chronometer and current position and any new chart drawn accordingly conformed to the Greenwich meridian. France initially used their own 0 degree meridian which passed through Paris, but despite using this in all other cartography simply translated The Royal Observatory tables based on Greenwich in the *Connaissance de Temps,* which simply became the French equivalent in translation of the British *Nautical Almanac.* In 1884, an international conference in Washington USA with representatives from twenty-six countries confirmed Greenwich as the prime 0 degree meridian of the world in reference to international

cartography and time-keeping ; as time-zones varied, the official 'Day' by which these are all arranged began at Greenwich and the International Date-Line became 180 degrees Longitude east-west, exactly on the other side of the globe from 0 degrees Greenwich and running from pole to pole across - roughly - the centre of the Pacific Ocean. France however disagreed with the conference vote and continued to use the Paris meridian (which is just over two degrees east of Greenwich) until 1911, but they still obstinately refused to refer to Greenwich Mean Time and used as a time reference *'Paris Mean Time retarded by nine minutes twenty-one seconds'* - but for every other nation from 1884, the world's clocks and watches everywhere were officially set to the time according to 0 degree meridian, 'Greenwich Mean Time'. Various visual signals to confirm Greenwich Mean Time had before 1884 already been instigated at each port ; according to Greenwich Mean Time, either a gun being fired or a ball dropping down a pole at exactly 1pm or 1300 hours within sight of ships anchored in the harbour so all chronometers - and pocket watches - could be checked by this signal and reset if necessary. Greenwich, London used a time-ball on a pole from 1833 ; every telescope aboard every ship in The Pool of London was trained on it from 12.55pm on that date. As every schoolboy knows, the time reference term AM means *'ante-meridian'* and PM means *'post-meridian'* based on the sun approaching or having crossed the 0 degree meridian. Modern recreated historical navigators using sundials or quadrants sometimes fall foul of the 'British Summertime' adjustment but fixing position in degrees remains constant if you don't adjust your watch from Greenwich Mean Time and use astronomy to set noon at your present position - or you could be around 30 miles out of position east-west for example on the latitude of Central London where one minute of longitudinal arc equals1.4 statute miles.

Latitude Calculations used in 18th Century Navigation

The distance between the 'parallels' of latitude shown on a chart remain constant north-south. The equally-spaced parallels of 15 degrees of latitude north-south of the equator stay parallel to each other as they approach the poles whereas the distance east-west between longitudinal meridians shrinks. At the equator the sun, moon and planets pass almost directly overhead ; the two other famous parallels *The Tropic of Cancer* and *The Tropic of Capricorn* mark the northern and southern boundaries of the sun's progress north-south over a year. The circumference of the globe had increasingly been more accurately measured from the return to Europe of Columbus in 1495. By 1669, the latest calculation showed the diameter of the earth as the equivalent of 12,554 kilometres (very close to the present-day measurement of 12,756 kilometres) giving a circumference of 24,902 miles. In 1669 it was still generally believed the earth was a perfect sphere though several astronomers including Sir Isaac Newton had previously stated that it wasn't, discovering the planet was compressed at the poles and describing the globe as an *'oblate spheroid'.* The planet's 'pear-shape' was finally confirmed and acknowledged in 1740 requiring to eliminate the previous error by yet another astronomical and navigational adjustment to all published tables, maps and charts. By measuring the angle of the sun above the horizon using a Fore-staff or a Davis Quadrant the latitude can be measured by an observer by subtracting the angle of altitude from 90 degrees with an extra calculation of a proportional seasonal variation as the sun gets higher or lower towards and away from midsummer. In England, the length of daylight hours widely varies seasonally between midwinter and midsummer with the halfway point between each marked by the vernal or autumnal 'Equinox', a latin term meaning *'equal night'.* The same reason why in the Arctic and Antarctic in summertime, the sun never actually sets for half a year - it just drops in the sky to skim the horizon before rising again - so technically shines at night giving the Arctic and Lapland the popular nickname in some quarters as 'The Land of the Midnight Sun'. Using a Hadley Quadrant, the observer would adjust the instrument's slides until looking through the eye-piece the bottom of the sun at noon was at the same level as the edge of the horizon then read off the measurement on the engraved scale of the instrument. For example, at sea-level at noon on the Equinox of 21st March or 21st September if the angle of the altitude of the sun is 50 degrees, you would then subtract 50 from the set 90 degree right-angle to get 40 degrees and that is your latitude north of the Equator at your position - but on any other day before or after the Equinox, a seasonal proportional calculation would have to be made for adjustment because of the sun's declination. By 1670, with the introduction and availability of various design developments of the Davis Quadrant often (referred to as a 'back-staff' and at that time still used in conjunction with the forestaff) and the yearly printed tables explaining methods of calculation, the calculating of latitude by a navigator or master's mate at sea had become comparatively simple to compute, though not allowing for magnetic variation in the compass and miscalculation of 'leeway' often caused a compound error. East-west global positioning - Longitude - remained the problem to be cracked as Dutch, French and English navigators and explorers attempted to expand their voyages and reach new destinations over distant and dangerous seas to establish new trading posts and new accurate charts based on their established ventures in the West Indies, Brazil, Africa, the Spice Islands, India, China and Japan. In the same year, the principle of Longitudinal measurement was fully understood : since the earth turns continually on an axis, there is nothing in the sky visible from one longitude that is not visible in the course of twenty-four hours from any other longitude - observers at different points on the globe see the same sun, moon and stars but at different times of the day. The earth is the original clock and decrees longitude can be determined through these time differences using two methods : the time of a celestial event observed (such as an eclipse) at a known longitude can be compared with the time for the same event observed at the location for which longitude is sought or by the navigator keeping track of the time at a known longitude (such as a home port like London) and comparing this to local time. In both cases, local ship-time has to be computed and compared by the navigator using the sun to show local Noon. The need for an accurate clock which was sea-worthy became the prime desire for all trading nations but this did not really begin to develop in England until the late 1730's with John Harrison's first timepieces. Until that year, astronomical methods prevailed using the moons of Jupiter using printed tables published in France in 1688 - but so difficult to do at sea as to be almost impossible. Even on land, accuracy to within a degree (60 statute miles in distance at the latitude of France) was beyond all but the most studious observer. A previous tabular work had been published in Holland in 1665 and copied into English ; described as a practical handbook for mariners, it offered ways of calculating local time at sea by timed observations of the sun but was also interpreted a sales attempt by the author at selling his pendulum clock, which had been proved to be somewhat lacking during sea-trials : *"... no certainty could be had from pendulum clocks for the longitudes as they never hung perpendicular and consequently the checks were false ; all kinds of motion upward and downward would alter the vibrations of them and any lateral motion would produce yet a greater alteration."*

With the development - and affordability - of powerful telescopes, all sorts of printed tables of accurate *'ephemerides'* (regularly printed tables giving daily sun declination, positions of stars, eclipses, high and low tides etc) for a given year enabling local time, latitude and longitude calculated by a navigator or observer became increasingly available from the 1670's, especially *The Royal Almanack* in England from its first publication in 1674 sponsored by the British Crown. In 1676, calculations by observers noted that all the maps and charts currently in circulation were in error by five degrees in the placing of longitudinal meridians between England and Jamaica. By 1688, printed tables emanating from the Royal Observatory 'shifted the blame' by including the reasoning that it was mariners themselves who were now responsible for the errors in calculating longitude by astronomy : *"I must confess that it is some part of my design to make our more knowing seamen ashamed of that refuge of ignorance, their idle and impudent assertion that the longitude is not to be found by offering them an expedient that will assuredly afford it if their ignorance, sloth, covetousness or ill-nature forbid them not to make use of what is proposed."* But - the method of using Jupiter's moons to accurately calculate longitude involved a large telescope, held steady enough to focus on the moving moon whilst the observer used a reliable clock to count the seconds before the small moon disappeared behind the planet and keep counting until the same moon appeared from behind Jupiter's shadow ; doing this regularly each day from the moving deck of a ship at sea - cloudy or wet conditions made observation impossible - was judged extremely unlikely by anyone with seafaring experience, some of whom wryly noted that this particular critic and all the other advocates of the 'moons of Jupiter' method of calculation had conducted all their observations around the planet always standing on dry land and one experienced astronomer attempting this method using a powerful telescope stated that even the tremors of his heartbeat caused him to lose sight of the tiny moons !

A squadron of warships in the tropics is 'becalmed' as the wind fails. Despite the lack of wind their ships would still change position by drifting with the tide and current - the large crews carried by warships would enable all the ships boats to be launched to tow the ships out of danger if the ships drifted too close to the reefs around the islands seen on the horizon in the background. The pennant flown from the masthead serves as a wind-gauge for those on deck.

All the simpler and reliable methods of calculating longitude still depended on accurate clocks or time-pieces ; the Royal Observatory was founded in 1675 to join the members of The Royal Society in trying to solve this and other associated problems, but especially making meticulous observations checking and improving the accuracy of the printed tables. In 1725, through the great need due to vastly expanding trade opportunities for a reliably accurate and sea-worthy clock had been reinforced by a maximum £20,000 reward from the British Parliament in 1714 for the successful designer, eminent astrologers and scientists such as Sir Isaac Newton were still writing to the Admiralty : *"The Longitude will scarce be found at sea without pursuing those methods by which it may be found on land. And those methods are hitherto only two ; one by the motion of the moon, the other by that of the nearest moon of Jupiter. A clock can only keep Time, it cannot find Location : I have told you oftener than once that Longitude is not to be found by clock-work alone. Nothing but astronomy is sufficient for this purpose."* It is interesting to note that despite almost continuous hostility between the three most boldest (or avaricious) trading nations at that time - Holland, England and France - during the last quarter of the 17th Century and the first quarter of the 18th Century the scale of international co-operation between the scientists of these three nations in trying to ensure a vital solution to accurate navigation through mutually-exchanged private astronomical, hydrographic and cartographic findings and initiatives.

The 'Time Machine' 1740 - 1800

As previously mentioned, the £20,000 reward offered by Parliament in 1714 resulted in many proposals sent to The Board of Longitude, but most of them were wholly unworkable. By the year 1740, only two real contenders for this reward were apparent : the 'lunars' aspect using the new Hadley Quadrant or Octant favoured by the astronomy faction - if one of the astronomers could crack the mathematical problems associated with it and present the information in a tabular form to be published for use by mariners - with the other option being the creation of a reliable and sea-worthy mechanical clock : an aspect at this time pursued only by one man, John Harrison. Latitude is fixed by the laws of nature ; but Harrison believed - rightly as it turned out - that Longitude is governed only by Time. As the earth revolves, taking 24 hours to complete a 360 degree turn, one hour marks one twenty-fourth of the distance ; or 15 degrees after the correct width of the globe had been determined. Each day at dawn or noon, each hours difference in the position of the sun before or after the recorded time of the same event at a home port's fixed time by local time gives a time difference ; for every hour difference the observer is 15 degrees west or east of the home port. One degree of longitude equals 4 minutes of Time on the equator but north or south of the equator, distance shrinks proportionally - a following calculation by the observer transformed the difference in degrees into distance - nautical miles - bearing in mind that the difference between the meridians at the equator is 68 nautical miles and shrinks to almost nothing at the Poles. Even if the observer possessed reliable clocks, multiplying a difference in hours by fifteen degrees only gave an approximation of the longitude ; he would then have to divide the number of minutes and seconds by four to convert the time reading to minutes and seconds of arc to fix his actual position. But - the weather aspect still meant that on a cloudy day no calculation based on the sun could be made and in place of an 'observation' a calculation by 'dead reckoning' by course and speed would be made, with the ships estimated position confirmed or adjusted through calculation by the observer at the next weather permitting 'noon sight'. The concept and mathematics of Latitude and Longitude could be and were taught for use by potential navigators - only more reliable instruments required to be developed to make the job easier for them at sea. By the late 1750's the 'lunars' aspect through many years of observations and calculations had reached a practicable point, though the astronomical mathematics required still left many navigators scratching their heads over a calculation that took an average of four hours to undertake - including several necessary checks and re-checks - as the altitude of various heavenly bodies had to be observed and measured, the angular distances between them also measured, adjustment made for the refraction of light at the horizon (making the observed objects appear higher than they are), an adjustment made for 'lunar parallax' as the associated printed tables were formulated from the centre of the earth, not at sea-level or the height of the observer above that. The Board of Longitude granted that there were difficulties and potential human error but by their very nature these difficulties once mastered in principle ensured accuracy. In fact, many navigators continued to make the same mistakes concerning longitude and magnetic variation that previous mariners always had, often resulting in serious delay and loss of life.

The Pacific Ocean covers over 64 million square miles ; if you compare the Pacific to the Atlantic Ocean, it is over twice the size and contains thousands of islands of various sizes. Considering the ratio of both time and space, the Pacific could easily contain the sum of the circumferences of all the continents on the planet with a large space left over. By the year 1600, Dutch ships - seemingly more able to apply improvements in navigation - had eclipsed the voyages of earlier Portuguese and Spanish explorers and penetrated into the West Pacific, founding their trading base at Batavia (Java) in 1618. In two voyages in the 1640's, a Dutch ship commanded by Abel Janszoon Tasman from established trading bases in the Spice Islands and colonies on New Guinea and Timor had sailed south and 'touched' *Terra Australis Incognito* - the continent which was then termed 'New Holland' but would later become Australia. Part of the coasts of *Van Diemen's Land* (now Tasmania) and New Zealand had also been seen by Tasman, but offshore dangers and especially *The Great Barrier Reef* meant that the eastern coast of Australia had not been visited and although Tonga and Fiji had also been sighted, the vast sea area from that point to Cape Horn remained uncharted and was largely unknown. By 1690 the dangerous shoals and reefs around the western coast of Australia had been negotiated by an English privateer and a Royal Navy warship and boats been able to land to explore the coast, resulting in the continent being largely written-off as a barren wasteland having nothing to offer and occupied by tiny groups of unfriendly and uninterested natives. The maritime chronometer created the opportunity for this vast ocean to be explored and accurately charted - but the necessary length of prospective 'voyages of discovery' still entailed great risk and a danger to the crew from scurvy. Finding 'stepping-stone' islands which could provide fresh food and water would remedy this problem - if there were any - hence any exploration depended on both finding and being able to accurately chart them. James Cook set out from England twice (in 1768 aboard *Bark Endeavour* and in 1772 with HMS *Resolution* and HMS *Adventure*) on his 'voyages of discovery' and successfully achieved this aim. Cook used the 'lunars' method through an Octant to compute longitude on his first voyage - but returned with very accurate charts - but was able to take a marine chronometer with him on his second to confirm that New Guinea was not part of the continent of Australia and begin the charting of the treacherous Endeavour Strait (now the *Torres Strait*) seen by the British Admiralty as their future *'gateway to the Pacific'*. John Harrison's H-4 chronometer was the result of thirty years work : after successful sea-trials in 1761 - where this accurate chronometer gained the complete approbation of the captain and master - a copy of H4 sailed with Captain James Cook on his second 'voyage of discovery' in 1772. In the intervening period, despite proof in several forms including signed affidavits from respected navigators and masters, the Board of Longitude had still not conceded that Harrison-based maritime chronometers had finally cracked the longitude problem mainly through their members having a 'vested interest' as the Board was made up primarily of astronomers who would only concede that an accurate maritime chronometer only *supported* the 'lunars' method, and various other objections to *'highly expensive, very complicated and unreliable mechanical devices'* as opposed to *'God's creation of the heavens which were available to everybody'* were made by some members with a religious bias. When Captain Cook returned to England in 1775, his reports included several glowing references as to the accuracy and reliability of the chronometer method enabling both safe and reliable navigation and in conjunction with a development of the Octant, very accurate chart-work had been possible to undertake for future navigators. As H-4 itself was far too valuable to risk at sea in the conditions expected, Captain James Cook's timepiece was a copy of H-4 and made over a period of two and a half years by Larcum Kendall, a London watchmaker, hence named K-1. As Cook through proven use became to rely greatly upon this chronometer he naturally became very fond of it, using the reliable information it gave to make the first very accurate charts of the South Sea Islands. Back in England he sang the praises of the chronometer in all his reports and logs, stating Harrison had with no doubt cracked the longitude problem and as a result, Kendall declined to go into mass-production. James Cook took K-1 with him again on his third voyage in an attempt to find the reputed 'Northwest Passage' and gave rise to the sea-legend that at the very moment Cook was assassinated by local islanders near Hawaii in 1779, the K-1 chronometer aboard his ship at the time for no apparent reason suddenly and mysteriously stopped ticking ... !

The only real problem remaining after the creation of H-4 and the acceptance of the technology behind it was the vast difference in price for captains or masters of vessels faced with choosing the 'lunars' or the chronometer method - a top-quality Davis Quadrant, all the available printed navigational tables and almanacks and the price of a new Octant still cost a tiny fraction of the price of a marine chronometer, which remained far out of reach for navigators. The cheapest timepiece then available to order cost a vast sum for the two years labour it took to make one, including materials such as the fitting of optional diamonds or rubies which had been found to further increase the accuracy of a marine chronometer (comparative18th Century values of labour and material costs with today's values are notoriously difficult, but a present-day equivalent of the above price would likely be well over £25,000). The aim from 1775 was to get the price of a marine chronometer - through *'mass-production'* - down to an affordable £200, a challenge which was taken up by two English watchmakers. By 1805, the price of a mass-produced box-mounted maritime chronometer made by either Thomas Earnshaw or John Arnold had fallen from £120 in 1785 to between £65 and £80 each and accurate marine chronometers were now within reach of most commercial maritime trading enterprises - and especially The East India Company, at the time exploring the possibility of opening up new trading posts in the Orient - though a parsimonious Admiralty meant that Royal Navy officers still had to pay for a marine chronometer out of their own pocket until the year 1800. The popular spread of these cheap but reliable chronometers by 1791 is shown through commercially-printed logbooks for use aboard ship in that year including a column for recording *'Longitude by Chronometer'.* Increasing numbers of navigators added their approbation in terms of reliability and accuracy to the chronometer rather than the quadrant for longitude calculation as the quadrant relied on printed tables and the extensive calculations and adjustments entailing several potential human errors which though regularly taught and studied aboard ship still caused many a Midshipman to fail his Examination for Lieutenant. Reliable mass-produced marine chronometers quickly became so affordable that it was not uncommon for a merchant vessel or warship in the year 1800 to be carrying several privately-owned reasonably reliable chronometers in the hands of captain, lieutenant, master, master's mate (and even some boatswains) who all kept daily checks on each others observations, but there are several examples after the year 1800 of ships at sea - even equipped with chronometers - still missing their destination or being lost through incorrect positioning and their crews dying through shipwreck, exposure, thirst and scurvy and also many reports of ships simply disappearing, never to be seen again ...

This particular pocket chronometer was made in Germany in the early 19th Century and measuring 4 inches wide by ¾ inch deep is a hefty piece of miniature engineering, with a key-wound complicated movement able to show time, date, day, month and sun & mood phase. Such items are highly collectable and hence quite rare so their individual value means you don't see many chronometers used in period navigational displays outside a museum. Despite being well over 150 years old, this chronometer is still an accurate timekeeper and used by the author in his displays. By comparison, H-4 is silver-cased and a beautiful example of devotion and craftsmanship.

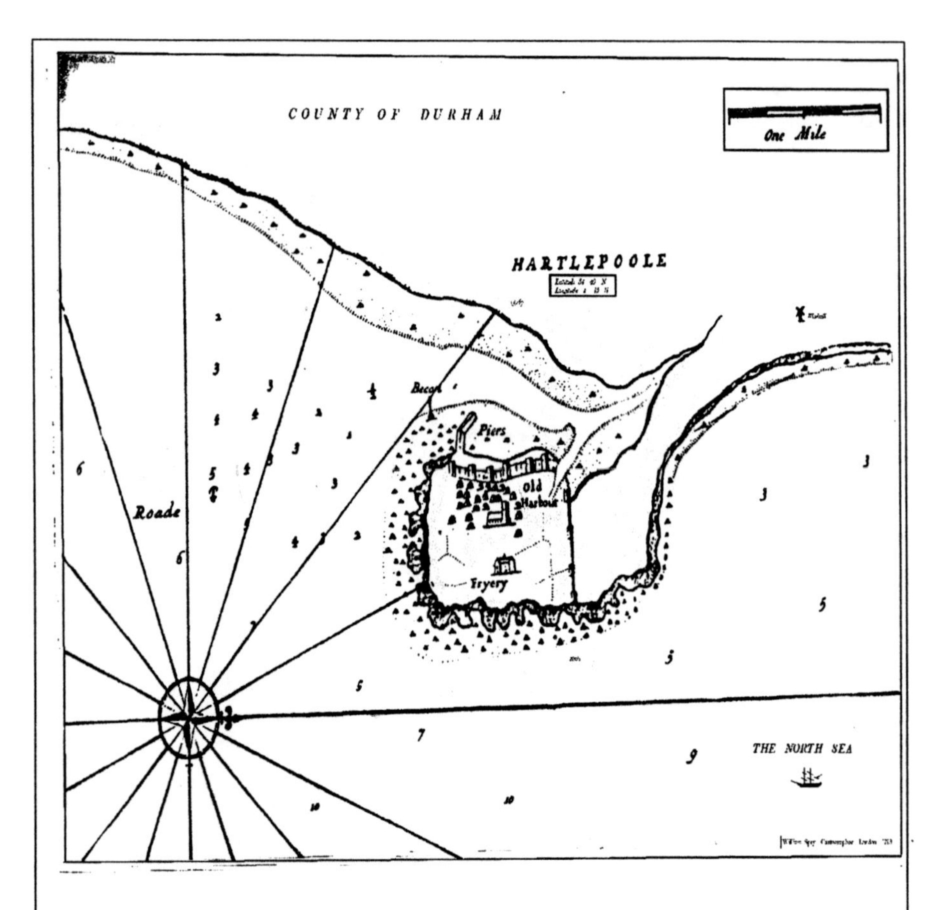

A small section of the original 18th Century chart has been enhanced here by the author for educational purposes. The small numbers dotted around the peninsula shown is the depth of water in fathoms at that place which would have been ascertained by use of the lead-line ; high-tide and low-tide water marks are also indicated. Not a good place for the ship shown on the chart to find itself in, as with a strong easterly wind there would be a great danger of being blown onto a rocky 'lee shore' and foundering there. A past but potential danger off such places were 'wreckers' : desperate men who in using false shore-lights deliberately encouraged a ship to run onto offshore rocks in bad weather at night in order to get their hands on both the wreck and the cargo. *(Courtesy of Hartlepool Maritime Experience)*

Try a simple exercise ...

Navigation in *'The Age of Sail'* does require some thought and skill to fully understand but becomes a little less daunting after the introduction of instruments aboard ship which can reliably compute both latitude and longitude (though nobody will ever eliminate bad weather). Try this method as an introduction - buy a world atlas or borrow one from the local library and start by finding England. Find the 0 degree prime meridian which passes through London and by placing your finger on Greenwich (London) you will then see almost every other place on the globe as being **EAST** or **WEST** of this line. Follow the meridian down to the Equator, where you will then have a **NORTH** hemisphere (above the equator) and a **SOUTH** hemisphere (below the equator). These are your first four navigational divisions : now find North and South America and lying midway between them is The Caribbean Sea and in the centre of the Caribbean Sea is the island of Jamaica. Is it Longitude *East* or *West* of the 0 prime meridian ? Is it Latitude *North* or *South* of the Equator ? Note that on your map, vertical meridians of longitude and horizontal parallels of latitude cross the Caribbean - follow the horizontal lines running **east-west** to the sides of the map where you will find the **latitude** scale, then do the same with the vertical lines running **north-south** where you will find the **longitude** scale. Narrow your reference point in the Caribbean Sea using these scales and find the position of another island lying roughly at Latitude 25 degrees **North**, Longitude 77 degrees 30 minutes **West**. Where the lines numbered above cross, you will see New Providence island (*Nassau*) which was once a favoured pirate harbour and base in the Caribbean. Now find a new position, stating once again from London and using the same scales as previously : Latitude 16 degrees **South**, Longitude 49 degrees 30 minutes **East**. Note that in doing so you will have travelled in the **opposite** direction from the previous position-finding and also moved **below** the equator into the southern hemisphere. The bay to the north-west at the position above is the old *'Pirate Round'* haunt - which was then named *Ranter Bay* - which is situated on the north-east coast of the island of Madagascar.

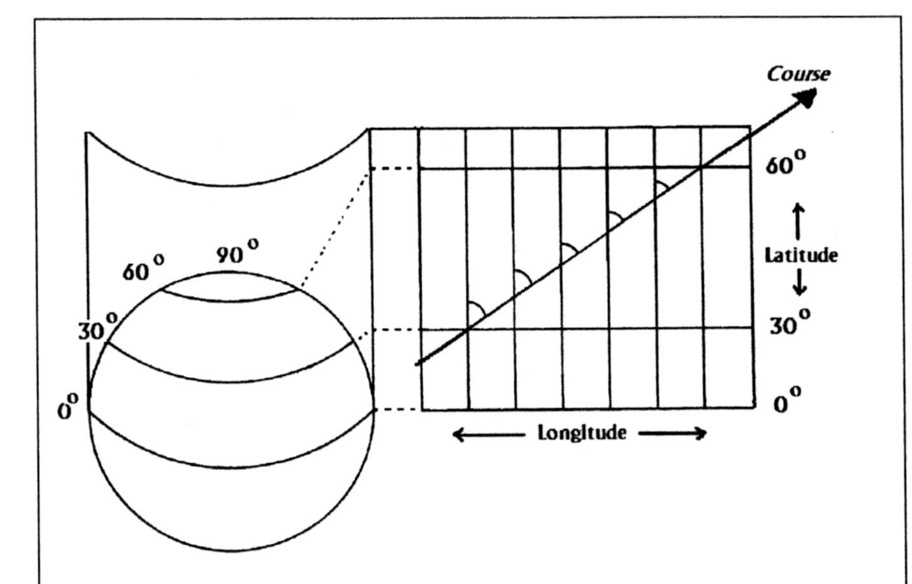

By the last quarter of the 18th Century, improved large-scale charts were available showing accurate parallels of latitude and meridians of longitude to enable a navigator to plot a long-distance course for a ship as a rhumb line on a Mercator projection. Measuring long-distance in nautical miles was taken into account by the expanding Latitude scale. This system (and the associated method shown in the following diagram) is often the one seen used aboard ship on feature films and television drama - whatever the historical period.

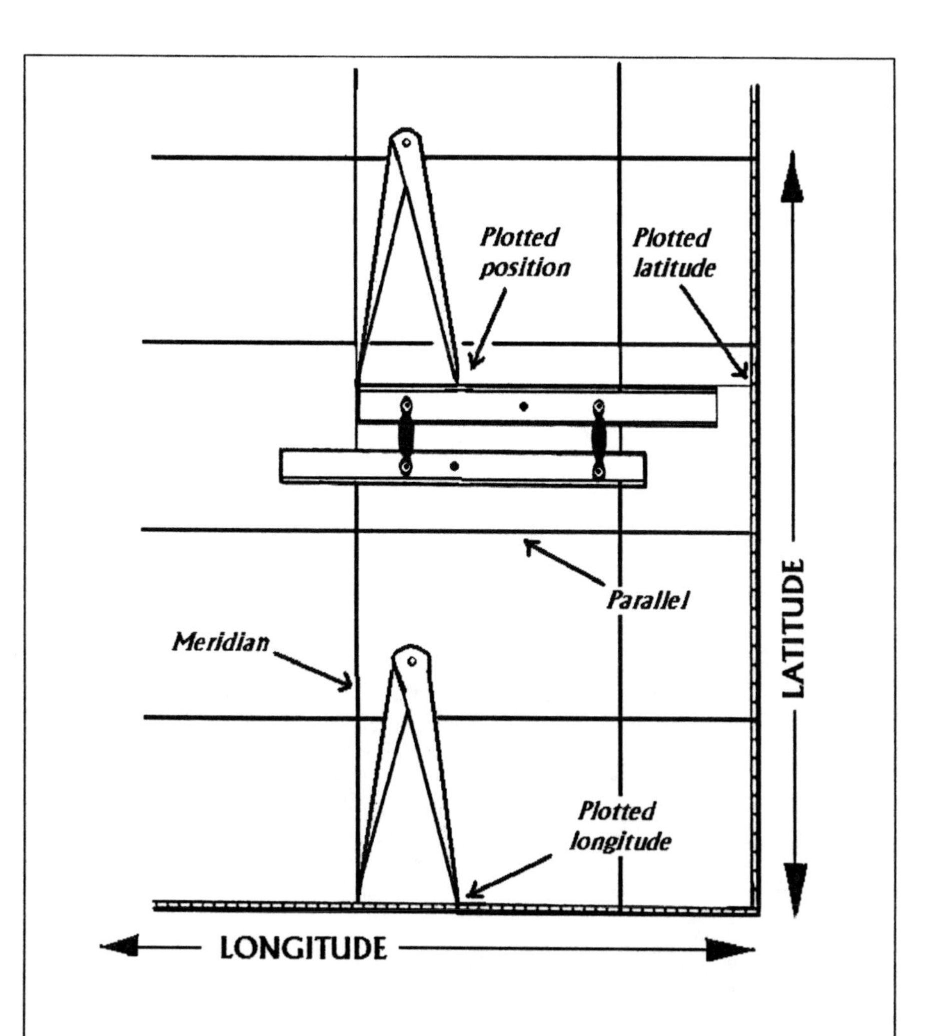

With a wide range of improved and available charts with a graduated scale along the edges showing accurate latitude and longitude, this method of plotting a ships position on a chart using an observed or calculated latitude and longitude became the norm using the instruments shown : a pencil, callipers and a 'parallel' rule - it is also handy to have a small lump of India-rubber to erase errors or unnecessary previous markings. Cheaper plastic instruments for beginners can be obtained from stationary retailers ; the above method is the system still in use today on modern charts in conjunction with telecommunicated information from Global Positioning Satellites.

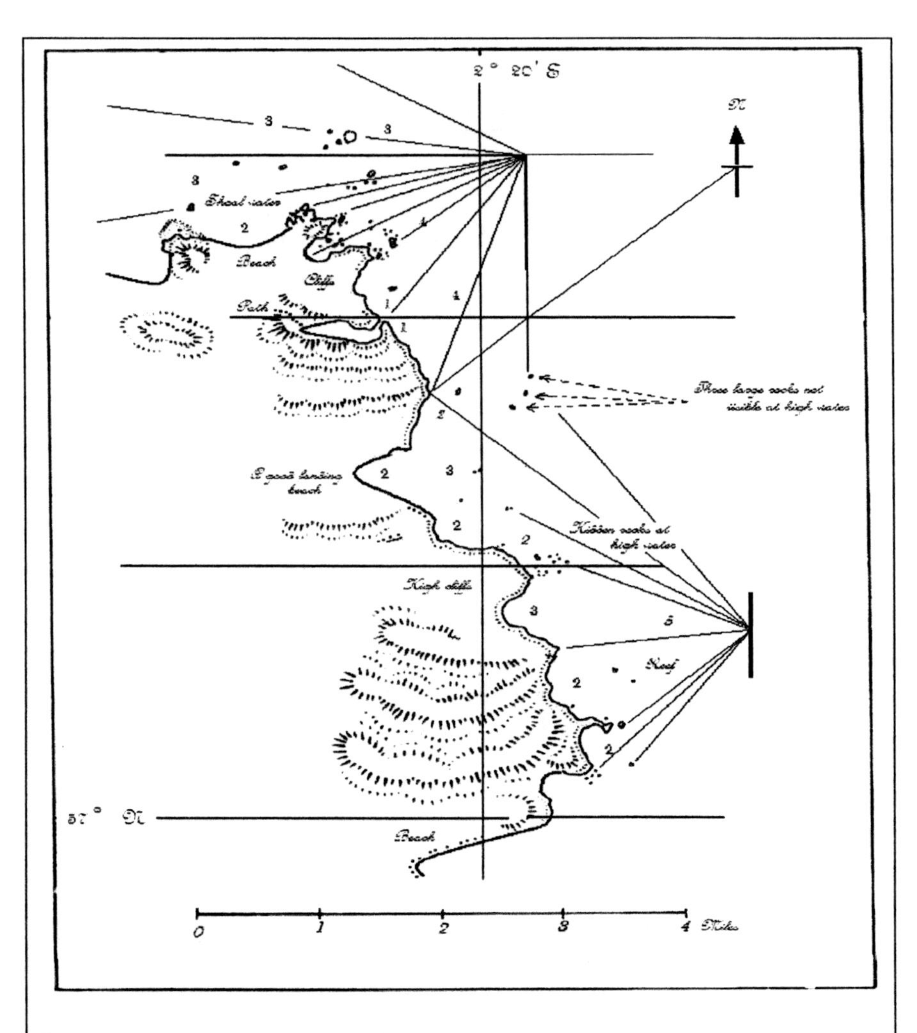

In comparison to the previous illustration, this is a facsimile late 18th Century hand-drawn survey by the author of part of a dangerous coastline which would be undertaken using an instrument known as a circumferentor, as mapping coasts in detail at that time went hand-in-hand with navigation. The radiating lines are those taken by offshore compass bearings. The 'map of an island' originally drawn by Robert Louis Stevenson was later used as an illustration to his famous novel *Treasure Island* shows similar hydrographic experience evident by the use of soundings, anchorages, rhumb lines and the inclusion of useful information ; the only navigational aspect omitted is the actual location of the island which would be given as a position in latitude and longitude but by using various historical clues contained in Stevensonian trivia the author deduced that it was intended to roughly lay Latitude 12 degrees North, Longitude 64 degrees 30 minutes West.

Now try another exercise, more 'in period' :
Take this exercise one step at a time. For the navigational exercises below in order to answer the above questions you will need basic instruments and a flat surface : a compass, parallel rule, dividers, a logbook (or just a sheet or two of notepaper) and a world map, a basic chart or a large Atlas. When *'laying off'* your course try and take into account the prevailing winds over the voyage, ocean currents and use an 'average' speed. For the purposes of this exercise, use the following simplified data : your ship can hold a maximum of 3000 units of stores or cargo - a single unit is either one cargo-capacity in the hold or a seaman's allowance of food and drink *per diem* (each seaman will always consume 7 units per week). Your ship has a crew of ten seamen and under full sail in a fair wind, should be able to make a respectable six knots - according to the previous logbooks your ship averages between 100 - 125 nautical miles *per diem.* For the purposes of the exercises, these voyages should be made within the general maritime parameters of the first quarter of the 18th Century as described in this book :

1. Give your home port of Bristol, England as a position of Latitude and Longitude ; compare it to your final destination of Port Royal (Kingston, Jamaica) given as a position of Latitude and Longitude.

2. From the log, what was the average speed of your ship given in knots per diem ?

3. 'Lay off'' a course for your ship from Port Royal (Kingston, Jamaica) to The Bay of Honduras for a speculative trading voyage. How many nautical miles is this prospective voyage ; and how many days might you be away at sea before your ship returned to Port Royal ?

4. After a hurricane, your ship is hired by the Governor of Jamaica to sail from Port Royal to Bermuda with a cargo of badly-needed stores. What is your maximum cargo capacity for this voyage ?

5. Your merchant ship is to sail on a 'Triangular Trade' round-trip voyage from Bristol, England. Note that stores are much cheaper to buy in Guinea (Africa) where you will first deliver 500 units of cheap metalware cargo before re-loading and filling your ship full to capacity with a varied cargo and sail for Port Royal. How many stores units will your ship require to sail from Bristol to reach Guinea ; and by revictualling the ship there, sail onto your next destination in Jamaica ?

THE 'ANSWERS' TO THESE EXERCISES ARE GIVEN IN APPENDIX 1

Epilogue

" Pieces of Eight ... ! Pieces of Eight ... ! "

The fabulous Spanish 'treasure-galleons' in sailing from Manila to Acapulco mentioned in the first part of this book had many problems of their own beyond the possibility of an enemy privateer or a lurking pirate ship waiting for the galleon to arrive in a weakened and desperate state at their destination. A voyage aboard one of these slow, usually over-crowded and over-laden vessels often took six months, with a high death-rate amongst the crew even on a voyage which reached the destination roughly on schedule. At least thirty of these mighty vessels were lost as a result of bad weather between the years 1550 and 1750 : in 1592, one of these galleons - being greatly over-due - was found by a *guardacosta* sent out to look for the ship found the vessel drifting off Cape Mendocino in California with all aboard the missing galleon having died from scurvy.

In trying to see why such established beliefs as the two mentioned in the first paragraph of the *Introduction* to this book may have come about, not only by looking at piracy in social terms as an escape from a hard and ill-paid profession but by attempting to ascertain an average early 18th Century death-rate amongst seamen, apart from the reputation of some captains for oppression and cruelty it did appear that one ship out of every ten that sailed may have suffered some form of premature end during a vessel's average lifetime in the first quarter of the 18th century. The mortality rate amongst seamen in the same period seemed to be on an overall par at least with Doctor Johnson's London even by leaving aside the problems with sudden fluctuations in trade and cheap gin, and mainly through overcrowding in the less desirable parts - which included Newgate Prison - causing regular bouts of general starvation and pestilence. All the long-haul ocean voyages recorded in print in the early 18th Century period seemed to be of the very boldest and hence the most risky - but the risk to the crews was not so much from a bullet, cannonball or a sword thrust in the fighting to obtain plunder, but in being exposed to long periods at sea, starvation and the sudden virulent tropical diseases in exotic locations such as *'Yellow Jack'* - a terrible West Indian fever which killed far more seamen and soldiers than any storm, bullet, cannonball or sword ever did - with a percentage mortality rate seemingly on a ratio with the increasing miles sailed (and also in the case of late 17th Century buccaneers, miles 'marched' overland through jungle or rain-forest). Admiral Anson's voyage into The Great South Sea in 1739 - though he did manage to capture one of the Spanish treasure-galleons and navigate his ship over an eventual 'circumnavigation' - comes over as the overall worse-planned venture of them all in not applying what should have been valuable experience from many previous ventures and enduring overall total crew losses of 70%. Another statistic that arose during this research was that most seamen aboard a ship at sea at any one time during the early 18th Century fell into the category of being *'young, fit and healthy'* because of the demands of serving aboard a ship at sea but because of these same demands were physically spent well before they were aged forty - what happened to merchant seamen who were obliged to 'retire early' and remain ashore because of this infirmity before reaching this age remains unknown. Apart from a lucky few, most pirates appear to have expired well before that age through a combination of the same exposure to hard work and the elements, with the added hazard of being captured and hanged and even for survivors of both these, succumbing to the self-inflicted hazards of sexually-transmitted diseases, an indifferent diet and a regular excess of strong drink : as Pirate Captain Bartholomew Roberts is traditionally said to have remarked, a seaman's life was very hard work in return for poor pay, causing many seamen to risk all and turn pirate in *"choosing a short life - but a merry one"*. Comparative figures taken from period records offer an average of 13,000 seamen serving in The Royal Navy between the years 1716 and 1726, with an estimate of 5000 seamen said to have *'turned pirate'* over the same period - for various reasons - for some part of their short lives, giving an average during that same period of around 1300 pirates sailing the seas *'somewhere in the world'* each year. Between 1700 and 1740, the population of Doctor Johnson's London approached half a million ; using these figures, during that period 0.01% of the combined populations of Europe and North America would have been engaged at some point in piracy and if an 'average pirate crew' did number somewhere between eighty and one hundred, this makes the average number of pirate ships afloat in this period to be around 15 - so it's easy to see why they were so elusive. Thanks mainly to Daniel Defoe, the lives of the *'Most Notorious Pirates'* were well-publicised and equally well-read, but if you read by comparison the earlier period *memoirs* of buccaneers and privateers such as Exquemelin, Dampier, Ringrose, Wafer, Pitman and Rogers, the view given in their books of the adventurous life of most privateers and pirates of this era does seem to bear out Bartholomew Roberts' remark on the subject, but with not so much of the 'merry'. *"Pieces of Eight!"* - the famous squawk of Long John Silver's parrot in the immortal novel *Treasure Island :* a Spanish-silver *'piece of eight'* having about the purchasing-power of an English sterling-silver crown coin, five shillings : doesn't seem to echo what most pirates in reality seem to have ended up with any great quantity of - the bag of 500 stolen pieces-of-eight received by one pirate as his share of a fourteen-month voyage that he proudly decided to retired on was worth £125 ; in today's values about a years salary. If a man stayed off the bottle and was willing to work, a sum of that sort could set him up for life in a comfortable trade ashore - but traditionally, few pirates chose to do this and squandered all their loot ashore in a month or two's spree of gambling, whoring and drinking to excess before returning - penniless, bleary-eyed and chastised - back *'on the account'.* A thrifty pirate that didn't blow all his money and returned home instead would not choose to advertise where all his cash had suddenly appeared from and wouldn't obviously include the fact in his *memoirs.* Using period figures, between the years 1700 and 1725 pirates worldwide stole - in today's values - about £1000 million in *specie* and trade goods but apart from a relatively tiny amount of confiscated loot used to build *The William and Mary College* at Williamsburg, Virginia and Greenwich Hospital in London where most of this money disappeared to is anybody's guess !

At the beginning of the historical period dealt with in this book, a large square-rigged three-masted wooden warship had been for many years the most complicated man-made machine that existed on the planet, and remained so for many years to come. On a happy note, by December 1831 *HMS Beagle* with the young Charles Darwin aboard sailed on that *'voyage of discovery'* carrying twenty-two accurate maritime chronometers to assist navigation and cartography and by 1860, the Royal Navy possessed four times as many maritime chronometers as it had ships to place them aboard. By that year, maritime navigation had for some years past no longer been the sole concern of a mysterious and enigmatic *'Ships Artist'* but had become practical for all seamen. Today, the Moon is better charted than the depths of our oceans and seas, which remain the most unexplored regions of our planet. Buccaneers and pirates - though they were sometimes known to draw charts - never drew a map showing where they had left any 'buried treasure' : but the trick in maritime navigation during *The Golden Age of Piracy* is to always make sure that wherever you put the 'X' on your chart it really does mark the spot !

" Keep Your Luff ; and Plenty of Duff ... ! "

(See Appendix 3)

The author at sea at the helm of a topsail-schooner approaching landfall and enjoying ideal sailing weather - a warm, steady breeze on a quarter, a clear sky and excellent visibility. By removing the metal hatch-cover from the deck and the life-jacket from the helmsman, this could be a scene during the latter part of the18th Century.

APPENDIX 1

The 'Answers' to the Exercises

If the reader wishes to recreate historical voyages on paper at home, take a copy of *Basic Seamanship* and *Prevailing Trade Winds* diagram in this book and use the diagrams in the above to make two simple 'wind gauges' cut out of stiff card ; place them both on a corner of your chart-table for indication purposes in setting / maintaining a course. To add some variation in his recreated historical voyages, the author had someone occasionally raise or lower the wind strength for a period of time by throwing two differently-coloured dice *(i.e. by 1 to 6 knots lasting from 1 to 6 days ; or more, by doubling-up both throws).* Throwing two sixes meant a sudden violent storm and sometimes a gale force strength of sixty knots or more lasting for two or more days, sometimes obliging the ship to simply *'run before the wind under reefed topsails'* or *'lay to under bare poles'* or ride out the storm using stay-sails to hold the hull bows-on to the wind and waves, with 'leeway' causing obvious concern here if the storm struck when the ship was off a rocky coastline towards which the wind then blew the ship : as in this instance, if the ship got so close to dangerous rocks or a coastline, you could eventually try dropping the anchors - but if the anchors dragged in the storm, then your ship is doomed unless the storm abated or the wind changed direction.

To continue, a potential navigator having read the main text of this book will already appreciate that *none* of the answers can be definitive as navigation at sea in the early 18th Century was not an exact science : the speed of your ship will vary according to the wind and current strength and the set of the sails. Your course is not dictated by simply drawing straight lines on chart and giving orders to the helmsman to theoretically sail along them by way of a compass course and time, but is a constantly variable adjustment to take into account wind direction, currents, tides, avoiding obstacles - small islands, reefs and sandbars - and the location of the ports where you will be re-provisioning or delivering / collecting cargo. Several times every 24 hours before and after Noon - and during the night - you or your Mate will have recorded or corrected your course and position through navigational astronomy and by keeping a *'Dead Reckoning'* when the sun, moon and stars were obscured by cloud or rain. Having such variables you can only *estimate* in making allowances for any course corrections, variable speed and 'leeway'. A potential Early 18th Century navigator may have already noted by now that the greater the scale of the chart - one showing the entire Atlantic or Pacific Ocean for example - the more difficult it is to plot an accurate latitudinal & longitudinal position and measure any distance in nautical miles. As a general guide - taking an average from ships logs during this period - a square-rigged ship would take 14 -16 days to reach The Canary Islands ; and from there, 7 - 9 days to reach The Cape Verde Islands. From Cape Verde, an Atlantic crossing to the nearest point in South America would take your ship between 28 and 32 days. Allow at least 21 days in a port for necessary calls, unloading/loading, being granted access to local markets and victualling the ship.

Exercise 1. ***DISTORTION*** Bristol : *Latitude 51 ° 30' N / Longitude 2 ° 30' W* and Port Royal : *Latitude 15 ° 32' N / Longitude 76 ° 30' W.* Some period charts will accurately show points of reference, but other charts may not. In 1705, many trading destinations in the world still had not been expressed as a reliable position of latitude *and* longitude - and though you might know the latitudinal & longitudinal position of the port of *Whydah* and *The Cape of Good Hope* but none of the potential stopping-places in between these. Until longitude began to be reliably calculated through the use of a marine chronometer from around 1770, charts would only show the longitude of a position if a scientist had actually been there and performed the observation and calculations ashore and also chosen to publicise his findings. The positional markings at the edges of the chart show latitude & longitude only at that point ; note that 0 degrees Longitude passes through Greenwich, London but a straight line drawn between the two reference points on the chart because of distortion may not. In taking a period reference book with you on the voyage, as you can see from the text just because the author or a previous navigator gives a positional reference for a destination it doesn't mean that it is correct !

Exercise 2. ***SPEED*** For navigational purposes, a ship travels for 24 hours *per diem* between the 'noon sights'. The 'log' will have been cast several times through the day to check speed - from the data given your ship is fast, with an average mileage *per diem* of 4 to 5 knots indicating an experienced Master and crew, but the ship may sail even faster in a stiff breeze with a full sail-plan and the wind on a port or starboard quarter. Sailing south off the African coast to reach Guinea towards Cape Blanco would require a course change to WSW towards the Cape Verde Islands to avoid the notorious Anguin sandbar stretching out from modern Senegal for miles ; easily done with the reliable north-easterly trade winds - but if the course was corrected too early, this would add a week to the voyage if the ship was slowed by the Equatorial current running westwards ; too late changing course with the equally reliable south-easterly trade winds at this point could result in 'tacking' into the wind for weeks to avoid the dangers of being caught on a 'lee shore'. The Anguin sandbar and bad navigation caused one of the most infamous shipwrecks known - the wreck of the *Medusa* in 1816 - resulting in the painting now hanging in The Louvre in Paris commemorating the dire consequences for the survivors who were *'deliberately cast adrift on an open sea'* by the ships' officers and the resulting state scandal in France involving the accusations of cannibalism caused by wilful negligence and alleged corruption. Because of the wind and current, a ship sailing from England to North America direct could take over three times longer to reach it than the same vessel sailing direct from North America to England.

Exercise 3. ***CIRCUMSTANCES*** may force you to change course drastically - a sudden storm, unexpected damage to the ship, the manifestation of any sickness amongst the crew or the outbreak of war. An adjusted sliding scale each day of what food and drink you have aboard ; sailing in excess of when these stores are due to run out is dangerous to the crew and the ship. A *rendezvous* or a planned return could be set for a calendar date but this would be generally understood to be a certain measure of time before / after (and set in days, weeks, months or in one extreme case, even a year). The distance from Port Royal to Honduras is 620 nautical miles on a WSW course after clearing the port which at an average speed would take your ship over six days to reach - but how long you might remain there before returning to Port Royal is an 'unknown' - if circumstances dictated haste and if he had to, an 18th Century captain faced with this problem with the possibility of an adverse wind and current on the return voyage would probably estimate a return to Port Royal somewhere between six and eight weeks from his initial departure. In an extreme example, a homeward-bound ship arriving at a foreign port in 1727 was suddenly alerted to the fact that the nation to which the ship belonged had been at war for over four months with the nation that governed the port into which they were sailing, causing the ship to be seized upon entry and the crew thrown into prison ; the unfortunate aspect was that the war had actually ended a month prior to their imprisonment but news of this didn't arrived there until six weeks later. An experienced navigator always considers the risks and all the options and alternatives to a set course.

Exercise 4. The distance from Port Royal to Bermuda through the 'Windward Passage' is just over 1200 nautical miles - this would take your ship around 12 days to cover, given a fair wind ; your ten-man crew would consume 10 stores units *per diem* on the voyage, so your cargo space would be : 10 x 12 days = 120 stores *units* / 3000 units capacity minus 120 equals 2880 *units* - but it is dangerous to cut things fine when sailing over open seas and sensibly allowing for possible delay, your maximum capacity would probably be 2500 *units*. Ships were sometimes overloaded with cargo and passengers (including using the weather deck for storage) and this could lead to difficulties as the ship would become *'cranky'* and slower and create handling problems in bad weather. The hold can be filled to maximise profit, but allowance must be made for the needs of the crew and the ship where storage space is at a premium. A warship with a crew of hundreds might provision for three months at sea, which would include a mixture of items of stores such as spare sails and spars, ammunition for the great guns and several sizes of live animals. You can see that each week at sea, your ship's crew require *70 units* of food and drink ; for each month at sea, *280 units*. A *unit* would equate to so many gallons of beer, pints of rum, chunks of cheese and four-pound pieces of salt beef or pork - and associated vegetables such as peas for as long as they remained edible - as laid down in each seaman's daily ration allowance to be calculated and transformed into weight of stores by the Quartermaster or

Purser ; or the same *unit* becoming the size and weight of different cargoes (in terms of volume and weight, a cargo of Indian pepper or Chinese rice would be calculated far differently to a cargo of Shropshire ironware). Fresh food has a limited life aboard ship and it would be of no use cramming aboard a valuable cargo-commodity - such as sugar - when still three weeks sailing-time from the nearest land, the ship after dealing with a spate of sudden adverse wind ran out of food and potable water - as unfortunately happened in one sad case in 1718. The mid-year 'hurricane season' in the Caribbean might see a ship forced to wait in a safe harbour for well over a month for bad weather to pass with the frustrated master and crew watching their cargo slowly rot or risk setting sail and being badly damaged and even destroyed - as happened in several cases.

Exercise 5. The width of the Atlantic Ocean from Cape Blanco in Africa to the island of Martinique in the Caribbean Sea is roughly 3000 nautical miles ; many merchant vessels are regularly making the 20,000 mile round-trip *'Triangular Trade'* voyage from England to Africa then across the Atlantic to the West Indies or North America before returning to England, loading and unloading several different cargoes at each place. The speed and handling of the ship will change with the cargo - setting off from Bristol with luxury goods, cheap manufactures, iron goods and brassware for delivery in Africa, perhaps loading with slaves for the Caribbean Islands plantation owners ; for the return leg to England from the Caribbean via the New England colonies filling the hold with sugar, rum & tobacco. A single voyage if successfully timed with the cargo intact could make such a profit that it would both pay for the ship and leave a substantial dividend for the owner or investors. It does - of course - take *time* to enter port, unload & load cargo, visit the local markets and re-victual the ship with food and drink and replace any losses to the crew before setting sail again, but staying in port the Master *might* if circumstances seem favourable give the crew an advance on their wages and shore-leave or even permit them to sleep ashore over-night (as those crew ashore were then obliged to consume food and drink at their own expense whilst you pocket the cost of their shipboard allowances) meanwhile restocking the ship with cheap local produce, salting the meat down with part of a cargo of salt picked up from the Cape Verde Islands which your ship would have passed.

A typical ship in the questions above was the vessel built on the River Thames at Deptford in 1697 which arrived at London in 1703 at the end of two *'triangular trade'* voyages and the crew were paid off after two absences of two years each voyage. The notoriously congested port of London was always very busy and the captain had to wait aboard ship for sixteen days before the cargo of tobacco and rum was unloaded into a excise-bonded store and - after customs dues having been paid - the profit distributed to the owners before his own account was settled with the crew fully paid-off during the following month. This 'lucky' ship made six more voyages for its owner-consortium before being judged to be *'no longer seaworthy'* and was broken up at New York in 1718 ; calculated over it's working life at sea returning an astounding profit well over 3000% on the investment made to build the ship. As the reader will now appreciate, sailing a ship involved far more than simply casting off and plotting a straight line course from A to B for the helmsman to follow. Even a tiny miscalculation which brought the ship to a destination ten miles down-wind instead of up-wind could result in several more days being added to the voyage in the crew having to *'beat up against the wind'*. Records of the course / speed have to be maintained if out of sight of land for any length of time and vessels with large crews such as warships - and to a lesser degree, pirates - would require additional record-keeping to show how long the remaining stores aboard would last and permit the ship to remain at sea, as to run out of water especially in the Tropics might quickly result in a mutiny and to run out of suitable food inflicts a crippling and debilitating sickness such as scurvy and both could mean the total loss of the ship. But - crossing the Atlantic or Pacific Ocean or reaching the Orient was increasingly undertaken by many ships from the middle of the 17th Century and despite modern navigational aids, some of the navigational hazards they began to encounter in the 18th Century - especially for modern *'Tall Ships'* and ocean-going yachtsmen during long spells of very bad weather - still remain today.

APPENDIX 2

The Use of the Cross-staff or Fore-staff

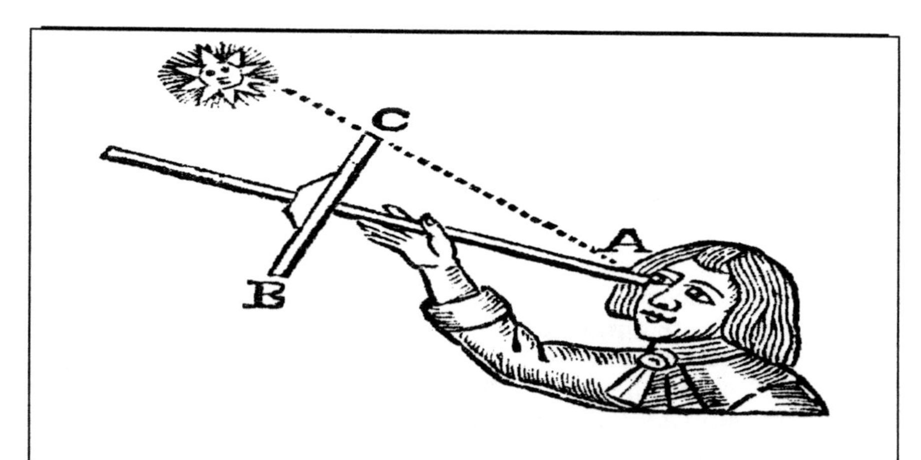

The Fore-staff is also referred to in period works as the Cross-staff or *Jacob's Staff.* Though this latter 17th Century illustration shows a navigator using the Fore-staff with the 'Sixty Cross' in making the Observation of the Sun having both his eyes open, the navigator would probably have whichever eye he wasn't using firmly closed when doing so and be using both hands to hold the staff steady enough and adjust the cross to make the observation. As previously stated, through the necessity of looking straight into the sun's glare the use of the Fore-staff made many experienced navigators blind, though it is hinted that some navigators placed a piece of coloured glass or a glass *'smoaked'* over a candle flame over their eye to try and limit this. From *Practical Navigation* by John Seller (London, 1669). *(Copyright National Maritime Museum)*

A restored original Forestaff showing all five constituent parts.
(Copyright National Maritime Museum, London)

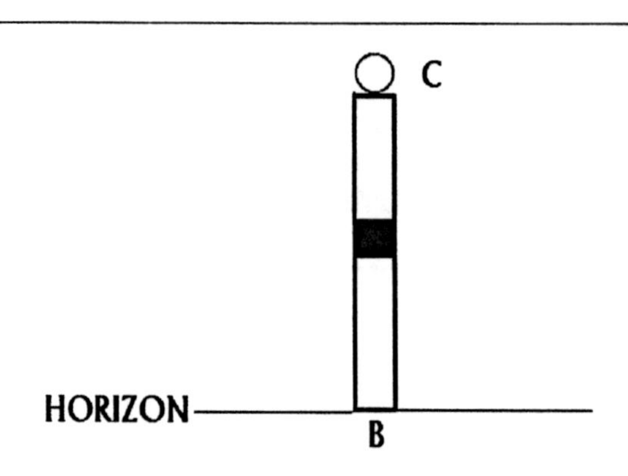

The view of the navigator in making the above observation. The cross is adjusted along the staff until the celestial body selected is balanced on one end of the vertical cross with the horizon aligned with the opposite end of the cross, when a reading of the angle between them can then be then taken pending which cross was used from one of the four scales on the horizontal staff. A slight daily variation would occur if the end of the staff was not placed in the exact position relative to the eye used for each successive observation. Using the sun, this observation is far more difficult due to the intensity of the light - so don't try recreating this method without adequate eye protection. By the inclusion in an early 18th Century navigational almanack of the following explanatory text, it shows that the Fore-staff was in popular use at that time.

"This instrument consists of a staff and four crosses, the first and shortest is called the Ten Cross and it belongs to that side of the staff which is numbered from about 3 degrees to 10 degrees. Sometimes the Thirty Cross, and the rest of the crosses are so made as that the breadth thereof serves instead of this Ten Cross.
The second cross is called the Thirty Cross and belongs to that side of the staff which is numbered from about 10 degrees to 30.
The third cross is called the Sixty Cross and belongs to that side of the staff which is numbered from about 20 to 60 degrees.
The fourth and last cross is called the Ninety Cross, and belongs to that side which is numbered from about 30 to 90 degrees.
This Staff is likewise numbered with the complement to 90 degrees ; at 10 stands 80, at 20 stands 70, at 30 stands 60 ;

and so the rest. The use of the instrument is to take the meridian altitude of the sun and the stars which is done as follows ; first, consider what the sun's greatest altitude will be that day and accordingly use the Cross most suitable - if the meridian altitude be judged to be under 10 degrees use the Ten Cross ; if between 10 and 30, the Thirty Cross ; if between 30 and 60, the Sixty Cross ; if between 60 and 90, the Ninety Cross (which is seldom used).
Having put the Cross onto the Staff, place the flat end of the Staff at A, to the outside of the eye, as near as maybe without hindering the sight ; thus, the Face being toward the Sun or Star, hold the Cross upright ; then look at the upper end of the Cross at C for the Sun or Star, and at the lower end at B for the Horizon. If the Sea appear rather than the Horizon, remove the Cross a little further from the eye, but if Sky appear instead of the Horizon, remove the Cross a little further toward the eye until the Sun or Star appear at the upper end and the Horizon at the lower end ; which when they do, then upon the side of the Staff belonging to the Cross used in Observation will be found the degrees and minutes of the Altitude of the Sun or Star. But the greatest Altitude being what is required, Observation must be continued as judgement shall direct, until the Sun or the Star be at its highest ; and as the Sun or Star rises, the Sky will appear instead of the Horizon, but when the Sun or Star is past the Meridian and begins to fall, the Sea will appear instead of the Horizon, and then is the Observation finished and upon the side of the Staff proper to the Cross used are to be found the degrees and minutes of the Sun's Meridian Altitude ; which subtracted from 90 degrees gives the complement of the Altitude or it may be taken off the Staff at once (the Staff being numbered with the Complement as is showed before) with which to proceed in finding the Latitude of the place, observe the Rules and Directions foregoing. "

The two methods of the *'Dutch Fashion'* in converting the cross-staff principle to 'back-sight'. In both, the cross is adjusted along the staff as previously described :

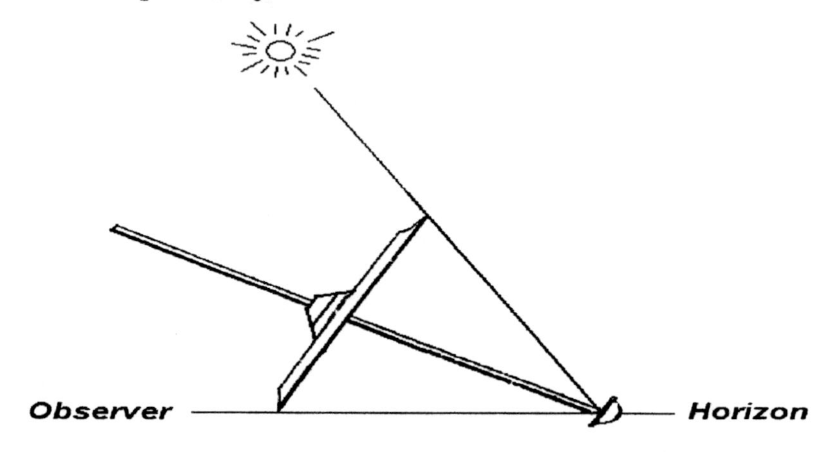

Above : Cross-staff converted by the fitting of a slotted horizon vane and using the high end of the cross as a shadow-vane

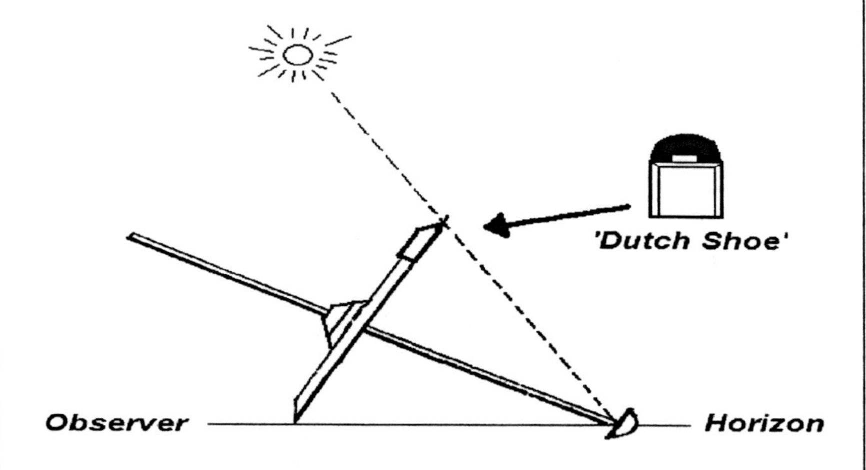

Above : converted Cross-staff using a *'Dutch Shoe'* ; the wooden shoe fits onto the high end of the cross selected, for sunlight to shine through the aperture in the disc at the top of the shoe and appear as a bright line on the horizon vane to improve the observation.

Appendix 3

“ Keep your Luff - and plenty of Duff ... ! ”

Advice from 'Long John' Silver to the helmsman when Quartermaster late off the *Cassandra* before he lost his leg aboard the pirate ship *Walrus* ; seaman's parlance for sailing as close to the wind as you possibly can in order to save time and gain an advantage in order to reap the benefits. By way of explanation, the ships *'luff'* is kept in sailing as close to the wind as possible to hold a set course towards the destination, though the ships speed would be reduced. Either a 'backing' or a 'veering' wind could comprise this course but satisfaction in doing so by the helmsman or navigator could be rewarded by the ships' cook in a serving of *'Plum Duff'*. This was a very popular 'treat' serving to mariners, welcomed at least once a week aboard an 18th Century ship and two genuine period recipes used aboard ship and since recreated by the author is as follows ; but note the alternative recipe given is based on having milk and eggs - which would apply if the ships' chickens and goats hadn't been devoured - but both of these recipes would raise the spirits of a hard-working ship crew and also offer good belly-ballast for a Navigator !

'Rough Duff' from a recipe (without using eggs) said to date back to 1692 : use 1 pound of plain flour, 1 pound of beef suet, two handfuls of currants, raisins or sultanas, ½ pound of potatoes and ½ pound of carrots, brown sugar and a pinch of mixed spice. Scrape the outer skin off the potatoes and carrots, grate them (or cut into small pieces and slice these up as thin as you can), mix in the other ingredients. Put into a greased bowl and cover, tie up in a cloth and suspend in an iron pot to boil for 3 hours. Remove and quickly slice into seaman-sized portions and pour neat 'navy' rum onto each slice whilst it is still hot. To get the best effect, the aroma from the hot rum should be slowly breathed in through the nose by the seaman before he begins to eat *(* see footnote below)*.

'Rich Duff' from a recipe (using eggs) of 1714 : 1.1/2 pounds of finely shredded suet, brown sugar 1 pound, raisins 1 pound, plums 1 pound (stoned and sliced up), lemon or orange peel 12 ounces, a teaspoonful of mixed

spice, half a grated nutmeg, 2 teaspoonfuls of salt, breadcrumbs 1 pound, plain flour 1 pound, eggs 1 pound (weighed in the shells), milk ½ pint and a wineglassful of brandy. Mix all the ingredients in a bowl, adding the well-beaten eggs ; add the milk and spirit, and stand aside for 12 hours in a cool place. Separate the dough into three 1 pound portions and place each in a pot, cover and boil in a large iron pot for 8 hours. If you wish to store a portion of this 'duff' for seamen coming 'off-watch' or next day, before serving boil that portion again as above for 2 hours. Quickly cut the 'duff' into suitable seaman-sized portions or slices whilst still hot and pour neat 'navy' rum onto each piece just before serving.

If you are shipwrecked ashore and only possess one large pot, the duff dough can be cut up and rolled into balls weighing about a pound each, firmly tied up in clean doubled cloth using twine or cod-line and using the strands, these bags suspended from a suitable thickness of wooden stick in a cauldron of boiling water for two hours. If you have no lid for your pot, use a plate (or try and figure out some other sort of cover). You will need to exercise great care in removing the cooked *duff* from the pot when cooked and cutting open the bags. Any 'landsmen' aboard may wish to have custard served with their duff instead of rum.

** The aroma from 'hot buttered rum' is recommended when trying to bring a deceased seaman back to life.*

The author as an Early 18th Century Mariner aboard ship about to fire a swivel-gun as a salute to the end of a successful voyage ...

VOYAGING FURTHER

All of the examples quoted in this book of the hazards and difficulties in period navigation appear in the following works:

Early 18th Century published Journals

There are several works available on the subject written within the chosen historical period. The works recommended below have since been reprinted as edited works but may still be difficult to find outside using a reference library ; each contain specific navigational quotes and very useful background material ...

The Buccaneers of America — *Alexander Exquemelin*
A New Voyage Around the World — *William Dampier*
A Cruising Voyage Around the World — *Woodes Rogers*
A Voyage around the World — *George Shelvocke*
A Voyage around the World — *William Funnell*

... not forgetting the classic source on period pirates : A General History of the Robberies and Murders of the Most Notorious Pirates *by 'Captain Johnson' ... but the author of this work could have been Daniel Defoe.*

Recent works

All these deal in part with the heritage of maritime navigational techniques and also offer very useful background information :

The Sinbad Voyage (1982) & The Jason Voyage (1985) by Tim Severin
The Wooden World by N A M Rodger (1986)
Longitude by Dava Sobel (1996)
The Prize of all the Oceans by Glyn Williams (1999)
The Ship : Retracing Cook's Endeavour *Voyage by Simon Baker (2002)*

Thanks & Acknowledgements
are given by the author to :

The National Maritime Museum, London - especially Sarah Grove of The Picture Library - for finding and sending the images

Nicolàs de Hilster - skilled hydrographic surveyor and re-creator of period instruments, for suggestions and advice

Replica *HM Bark Endeavour* (Stockton-on-Tees)

The Hartlepool Maritime Experience

Sheppey Pirates shipmates in *The UK Pirate Brotherhood* who had a go on paper at some of the theoretical voyages

William Spry,
18th Century Man.

Take a journey into the 18th Century with him by reading .....

THE WORLD OF WILLIAM SPRY, ESQUIRE

by Richard Rutherford-Moore

With a foreword by Bernard Cornwell, author of SHARPE

Published by Heritage Books / Willow Bend Books USA & available on the internet from AMAZON

The author is now writing *'Dead Men Do Tell Tales'* - taking a detailed look at how a pirate ship was crewed, sailed and operated between the years 1690 - 1720

INTRODUCING

Richard Rutherford-Moore
(author, international historic tour guide, guest speaker/lecturer)

Richard has been interested in researching and recreating history from boyhood, being a long-term muzzle-loader and 'living history' interpreter. Due to his success in portraying a typical soldier of Wellington's Army for *English Heritage* he came to the notice of television producers and served from 1991-7 as the Military & Technical Adviser and Armourer on the enormously successful television drama series, *Sharpe*; in 1998 Richard went on to serve aboard a 19th century style square-rigged three-masted sailing ship as a crew member on the first television episodes of 'Hornblower'.

Richard can be seen on-screen as the popular 'Rifleman Moore' in many episodes of *Sharpe* (particularly *Sharpe's Battle,* part of which Richard penned) and as 'Captain Harvey of the Dockyard' in the *Hornblower* episode

'Examination for Lieutenant'. Richard has just returned from India after serving in his usual role (as above) both off- and on-camera in the new series of *Sharpe's Challenge*, the biggest television drama project in the world in 2005.

Richard regularly guided popular battlefield tours to Portugal, Spain, Belgium covering the period 1808-1815 in both a practical, tactical and strategical sense - but is said to be an excellent and evocative 'story-teller' too - and since 1997 has been taking tour parties to the Crimea to guide unique and detailed tours over the embittered landscape of the Crimean War of 1854-6. Richard led the battlefield tour to the Crimea in 2004 on the 150th Anniversary of *The Charge of the Light Brigade* and has contributed to two television documentaries on the subject.

Richard's knowledge, ability, outdoor skills, charm and affability are said to make him an ideal companion on these trips, scoring a consistent 100% on performance on customer evaluation surveys. Richard has advised and contributed to The History Channel, The Discovery Channel, BBC, Channel Four and Five and many other satellite and terrestrial production companies on historical subjects. Richard also presents unique talks on a variety of historical subjects.

Richard's most recent 'living history' presentations have concerned *Pirates !* His new book for 2007 is named *The Pirate Round* and examines in detail the relatively unknown and mysterious aspect of maritime navigation during the period 1690-1730. Richard also has enjoyed a great deal of success with a new popular illustrated talk named *The Truth about Pirates !*

Richard lives in an old haunted house in Sherwood Forest, England. Though he penned several books, his previous book with Heritage Books was *The World of William Spry, Esquire* which recreated the life of a 'Georgian Man 1740-1815', a rather dubious 18th century character he created in 1990 and well-known amongst period re-enactors in the UK for his knowledge and 'attention to detail'.

e-mail : armor1@ntlworld.com

Made in the USA
Charleston, SC
03 December 2009